Urban Sustainability

The Urban Sustainability Book Series is a valuable resource for sustainability and urban-related education and research. It offers an inter-disciplinary platform covering all four areas of practice, policy, education, research, and their nexus. The publications in this series are related to critical areas of sustainability, urban studies, planning, and urban geography.

This book series aims to put together cutting-edge research findings linked to the overarching field of urban sustainability. The scope and nature of the topic are broad and interdisciplinary and bring together various associated disciplines from sustainable development, environmental sciences, urbanism, etc. With many advanced research findings in the field, there is a need to put together various discussions and contributions on specific sustainability fields, covering a good range of topics on sustainable development, sustainable urbanism, and urban sustainability. Despite the broad range of issues, we note the importance of practical and policy-oriented directions, extending the literature and directions and pathways towards achieving urban sustainability.

The series will appeal to urbanists, geographers, planners, engineers, architects, governmental authorities, policymakers, researchers of all levels, and to all of those interested in a wide-ranging overview of urban sustainability and its associated fields. The series includes monographs and edited volumes, covering a range of topics under the urban sustainability topic, which can also be used for teaching materials.

Kai Wang

National Spatial Planning in China

Theoretical Approach and Applied Practice

Springer

Kai Wang
China Academy of Urban Planning
and Design
Beijing, China

ISSN 2731-6483 ISSN 2731-6491 (electronic)
Urban Sustainability
ISBN 978-981-97-7728-0 ISBN 978-981-97-7729-7 (eBook)
https://doi.org/10.1007/978-981-97-7729-7

This Springer imprint is published by the registered company Springer Nature Singapore Pte Ltd.
The registered company address is: 152 Beach Road, #21-01/04 Gateway East, Singapore 189721, Singapore

中规院
CAUPD

This book is published with the support of the China Academy of Urban Planning and Design. 本书由中国城市规划设计研究院资助出版.

Foreword I

Spatial planning plays a crucial role in the development of a country or region. China's modern spatial planning can be traced back to the late Qing Dynasty and the early Republic of China, starting with figures like Zhang Jian and Sun Yat-sen. It has been a journey of arduous exploration and twists and turns. Over the past 30 years of reform and opening up, China's urbanization has experienced rapid development, marked by significant progress and transformations, as well as substantial waste and destruction. Currently, we are facing both exceptional development opportunities and complex contradictions and challenges. To tackle these complex issues, we need to adopt a scientific theoretical framework and holistic thinking to systematically understand and address them. For spatial development issues, such as the strategy for developing the western region, various opinions exist, but without the establishment of territorial spatial planning, policies cannot be effectively implemented.

Wang Kai has been pursuing his doctoral degree at Tsinghua University since 2001. He is diligent in his studies and thoughtful in his thinking. During this period, he led the work on National Urban System Planning (2005–2020) and made significant contributions from conception to outcomes. I once encouraged him to build upon this foundation, break free from departmental constraints, and conduct research on national urban spatial planning from a territorial perspective. Now, it is evident that through his arduous and long-term endeavor, he has earned remarkable achievements.

Throughout this year, more than ten regional-level plans were successively introduced, which undoubtedly have significant implications for national and regional development. At this moment, Dr. Wang Kai's publication of this book based on his doctoral dissertation can attract attention from the academic community and certainly stimulate discussions among researchers, thereby promoting the progress of spatial planning practice in China. I often say that the science of human settlements is an applied discipline, and this is precisely where the unique value of this book lies.

Beijing, China
September 2010

Wu Liangyong

The original version of this book has been revised. Foreword II has been included.

Foreword II

Over the past decades, China has seen unprecedented economic progress and urbanization. However, outside of China little is known about the underlying theories, policies and strategies, which have contributed to make China the second-largest economy and the largest predominantly urban country in the world. It is not surprising that national spatial planning has played a pivotal role in this process. From its beginnings in the early 1900s to its current focus on sustainable development and the so-called "eco-civilization", national spatial planning in China has been characterized by a fascinating, though internationally rather unknown sequence of phases with varying features.

Although the share of Chinese publications in the international scientific planning community is constantly rising, there is still only a vague understanding of spatial planning in China in the outside world. This is partly due to the nature of publications, which often focus on spatial-data-based case study analyses and pay little attention to systemic background information. Moreover, it can be attributed also to differences in understanding and using planning terms in China and in other countries, which often lead to misunderstanding.

On this background, the book by Wang Kai about national spatial planning in China contributes to filling an enormous knowledge gap. It provides an informed view on the various challenges, which China has faced over the decades, and it delves into the different planning approaches, which have characterized China's development path. The book draws a connection from a retrospective to future perspectives of China's National Spatial Planning.

The author has long been engaged in urban and rural planning and research of the China Academy of Urban Planning and Design (CAUPD), and served as the organization's chief expert in many national R&D projects in the field of urbanization. Currently he is the President of CAUPD and one of the most masterminds behind China's way of orientating its spatial planning system.

The book is based on the author's dissertation at Tsinghua University, one of China's top universities, published in the Chinese language almost 15 years ago. The largely changed and updated version provides a timely and comprehensive overview over current national spatial planning in modern China, and traces its foundations back to the roots.

The book is a "must-read" for all those who want to know more about China's national spatial planning system and its background. One of the special strengths of the book is that it connects planning and development in China with international development theories and trends in urban and regional development. Moreover, it uses an internationally compatible terminology and it explains Chinese trends in an internationally understandable way. Thus, this book is a seminal publication about spatial planning and development in China. It will be a standard book in libraries all over the world. Moreover, it deserves to find its way into Chinese universities and their planning schools as an indispensable source of information for students and their instructors.

Xuzhou, China/Dresden, Germany
June 2024

Bernhard Müller

Preface

This book was developed based on my doctoral thesis undertaken at Tsinghua University from 2001 to 2006, a journey overseen with meticulous care by Prof. Wu Liangyong. Since my graduation from Tongji University in 1986, where I specialized in Architecture and Urban Planning, my career path has been intertwined with China Academy of Urban Planning and Design, engaging in diverse range of urban planning design and research activities. I have contributed to large-scale national, provincial, and capital city planning endeavors, along with projects in smaller towns and villages. In recent years, my involvement in numerous research initiatives spearheaded by the Ministry of Housing and Urban-Rural Development, the Ministry of Science and Technology, and the Chinese Academy of Engineering has broadened my insights into urbanization, urban systems, and urban continuity of large cities, offering me a fresh perspective on macro-level planning. Notably, my role as a key member of working group for the 2005 initiative of National Urban System Planning presented an opportunity to reflect on the significance and function of national spatial planning in the context of market economy by systematically taking stock of research findings of spatial planning at national level. Originally, I intended to explore the intricate relationship between capital and urban spaces in the Yangtze River Delta against the context of globalization in my doctoral thesis, with part of preliminary contents already completed. However, under the acute guidance of Prof. Wu, I redirected my focus toward the emerging domain of national spatial planning, a critical theme that awaits careful deliberations within contemporary Chinese urban and rural planning paradigms in the new era. Embracing his counsel, I embarked on an in-depth study in this area. Throughout the refinement of my paper, Prof. Wu's academic rigor, profound expertise, and acute insights proved instrumental to the successful completion of my dissertation. In recent years, many scholars have questioned the constantly released new plans for its lack of systematic analysis, with the rising number of state-approved regional plannings of various kinds. In retrospect, I have come to recognize the foresight of Prof. Wu as he prompted this shift in focus. While my original intention was to make a systemic revision in the context of new findings from the past two years before publication, some aspects of the revised edition fell short of my aspirations. Consequently, I opted for a comprehensive redraw and slight textual adjustments in preparation for publication, with plans for a future systematic revision when possible.

Under the guidance of various experts including Academician Zhou Ganzhi, Academician Lu Dadao, Prof. Zhou Yixing, Prof. Mao Qizhi, and Prof. Wu Weijia during my doctoral journey, their careful reviews have been instrumental. I was particularly moved by the meticulous feedback by Mr. Lu. I also extend my heartfelt appreciation to Prof. Zuo Chuan, Mr. Wu Tinghai, and other colleagues at the Institute of Architectural and Urban Studies, Tsinghua University, for their valuable input.

Continuing studies while managing demanding professional commitments may entail self-imposed hardships, yet I find this journey to acquire new knowledge and insights a gratifying endeavor. The saying that celebrates the learning of truth in the morning even with upcoming death in the evening is broadly speaking akin to this sentiment. My gratitude is extended to President Wang Jingxia and President Li Xiaojiang from China Academy of Urban Planning and Design for the valuable opportunity to learn. Special acknowledgment is reserved for School of Urban and Regional Planning at Cardiff University, wherein the guidance and support from Ms. Alison Brown, Mr. Jeremy Alden, and Mr. Yu Li during my study tour in the United Kingdom greatly enriched my academic journey. I express my thanks to my fellow colleagues Li Rong, Xu Ying, Li Hao, and others for their dedicated efforts in organizing and enhancing the visual content of this work. The unwavering support of my family throughout my academic pursuit also stands as a beacon of inspiration that motivates my scholarly endeavors.

Beijing, China
May 2010

Kai Wang

Preface to the English Edition

Since the publication of the Chinese version of the "National Spatial Planning Theory" in 2010, China's national spatial pattern has rapidly evolved, reflecting new dynamics in the urban spatial development. On the one hand, China's accession to the WTO and comprehensive openness has positively influenced the national spatial pattern over the past decade, leading to the emergence of urban agglomerations, metropolitan areas, and central cities. On the other hand, significant events such as the 2008 global financial crisis and the pandemic since 2020 have also had a significant impact on urban spatial distribution and organization. In light of this, I have further compiled relevant data indicators on China's urban spatial development over the past 20 years, examined the challenges faced in urban spatial development, and provided a new outlook for China's national spatial pattern in the future. The conclusions of this part of the study have been added as a new Chap. 1 to the original book.

Based on the understanding and research through recent years, I have continued to analyze the latest trends in sustainable development and, drawing from my over a decade of regional planning practice in China, further enriched and improved the theoretical and methodological framework of national spatial planning. I have summarized a "precise adaptation" technical system that aligns urban spatial distribution with ecological, safety, and cultural foundations, which will further guide regional planning in China. Relevant research results have been added in the new Chap. 6. Through the publication of this book in English, I hope to share the latest progress in regional planning research in China with regional planners and policymakers across the world and provide insights for exploring national spatial planning that is tailored to the urbanization realities of each country.

This book is a summary of my research and contemplation on regional planning throughout my more than 40 years of work at the China Academy of Urban Planning and Design. It is also one of the academic achievements of our institute's 70th anniversary. I would like to express my sincere gratitude to Academician Wu Zhiqiang and Academician Müller for their careful guidance in the publication of the English edition. I would also like to thank my colleagues at the China Academy of Urban Planning and Design, including Chen Ming, Xu Hui, Zhou Yajie, Fu Kai, Luo

Qianyi, Wang Ying, Zhang Danni, and Jia Pengfei, for their assistance and support in revising the manuscript. I want to particularly acknowledge the significant efforts put forth by Ms. Luo Qianyi in the translation and proofreading process. Furthermore, I extend special thanks to Director Li Yan and Senior Editor Wang Lei from Springer Nature (Beijing) for their active promotion of the publication of this book.

Beijing, China
February 2023

Kai Wang

About This Book

This book provides a systematic and in-depth study of China's urban spatial development issues from a land perspective, based on sciences of human settlements. By integrating theories such as globalization, spatial planning, and national intervention, it systematically reviews the theories of urban spatial development at the national level both domestically and internationally. Taking into account the influencing factors of population, industry, resources, ecological environment, and management mechanisms in China's urbanization and urban development, it presents theoretical and methodological approaches for urban spatial development planning in the era of globalization, with a focus on the construction of a favorable human settlement environment. This book is suitable for researchers in the field of sciences of human settlements, professionals in urban and regional planning. It also serves as a reference for scholars, and graduate students in disciplines such as urban planning, geography, architecture, and regional science at higher education institutions, as well as for government officials.

Contents

About the Author

Kai Wang, a national master of engineering survey and design, serves as the president of the China Academy of Urban Planning and Design. He holds a Ph.D. in Engineering from Tsinghua University and possesses the expertise and rank of a professor-level senior planner. He serves as a member of the Expert Committee on Human Settlement Environment of the Ministry of Housing and Urban-Rural Development (MOHURD), Chairman of the Urban and Rural Construction Planning Standards Committee of MOHURD, Vice Director of the China Urban Planning Society, Vice Director of the China Architecture Society, and a member of the International Society of City and Regional Planners (ISOCARP). He is an adjunct professor at Tsinghua University, Cardiff University in the UK, and Renmin University of China, where he also serves as a doctoral supervisor.

Kai Wang has been involved in numerous major national scientific research projects and is the chief expert of the Ministry of Science and Technology's major scientific and technological special projects in the field of urbanization from the 11th Five-Year Plan to the 14th Five-Year Plan. He has led the planning of national urban systems, the planning of the Beijing-Tianjin-Hebei urban-rural areas, and the urban system planning of provinces such as Xinjiang, Jiangxi, and Shanxi. He has also participated in and led the master planning of cities such as Beijing, Taiyuan, Haikou, and Qinhuangdao and has been in charge of the development strategies of cities such as Hangzhou, Ningbo, and Xiamen. He has published over 100 papers and authored several books, including National Spatial Planning, Trend and Evaluation of China's Urbanization, Classification and Distribution of China's City Clusters, 40 Years of New City New Districts in China: History, Evaluation and Prospects, and Knowledge and Practice in Planning. He has received awards such as the Third National Innovation Leadership Award, the Standard Master Award from the China Engineering Construction Standardization Association, and the Leading Talent Award from the China Urban Planning Society.

Chapter 1
A 20-Year Retrospective and Trend Analysis of China's National Spatial Development

Over the past two decades, China's comprehensive national strength has continuously increased, leading to significant changes in the national spatial pattern and profound impacts on the world economy and global urban systems. The country's share of global GDP has risen from 3.6 to 17.0%, making it the world's second-largest economy after the United States. The urban population has grown from 450 to 930 million, experiencing the largest and fastest urbanization process in world history. According to GaWC research, by 2020, China had 26 cities meeting global urban standards, accounting for 11.6% of the global total, thus becoming a vital component of the global urban system. Concurrently, China's urban spatial structure and morphology have undergone significant changes, giving rise to mega-city regions such as the Yangtze River Delta and the Guangdong-Hong Kong-Macao Greater Bay Area, as well as the emergence of distinctive urban agglomerations and metropolitan areas, while some regions and cities have faced challenges of decline.

Nevertheless, it is important to acknowledge that major events such as the 2008 global financial crisis and the ongoing COVID-19 pandemic since 2020 have severely impacted the global economic system and prompted critical reflections on previous development models. Like other countries worldwide, China needs to address challenges related to climate change, environmental resource protection, and achieving social equity, while also grappling with the complexities of spatial and social governance as a major power. These pressing issues require continuous exploration and research to identify solutions tailored to China's specific circumstances.

K. Wang, *National Spatial Planning in China*, Urban Sustainability,
https://doi.org/10.1007/978-981-97-7729-7_1

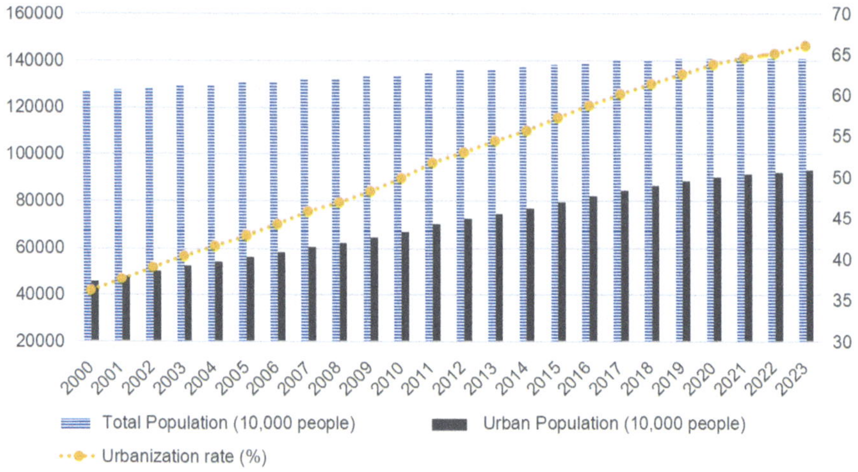

Fig. 1.1 Trends in China's total population and urbanization level (2000–2023). *Source* China Statistical Yearbook

1.1 Changes in China's Spatial Pattern Over the Past 20 Years

1.1.1 The Second Half of Rapid Urbanization Advancement

Overall, China has remained in a phase of rapid urbanization over the past two decades. The urbanization level has increased from 36.2% in 2000 to 66.2% in 2023, with an average annual growth of 1.3 percentage points. In 2011, the urbanization rate surpassed 50%, marking a historic milestone as the urban and town population exceeded the rural population for the first time. Since 2020, as the urbanization level continued to rise and economic growth slowed down, the pace of urbanization has slowed, with an average annual increase of less than 1 percentage point, entering the later stage of rapid advancement. The total population peaked at 1.42 billion in 2021.[1] According to the United Nations' World Population Prospects, it is projected that China's total population will decline to around 1.32 billion by 2050, and population aging and a declining birth rate will be long-term challenges faced by China (Fig. 1.1).

Population mobility has become increasingly active. In 2020, the scale of China's floating population reached 376 million.[2] This accounts for over one-fourth of the national population and is approximately three times the number in 2000, indicating

[1] The data on urban, rural, and total population, regional GDP, import and export data mentioned in this book do not include data from Hong Kong, Macao, and Taiwan.

[2] The floating population refers to individuals who are not included in the household registration system within the jurisdictional area. Intra-municipal floating population refers to individuals within a municipality or prefecture-level city who reside and register their household in different townships or streets.

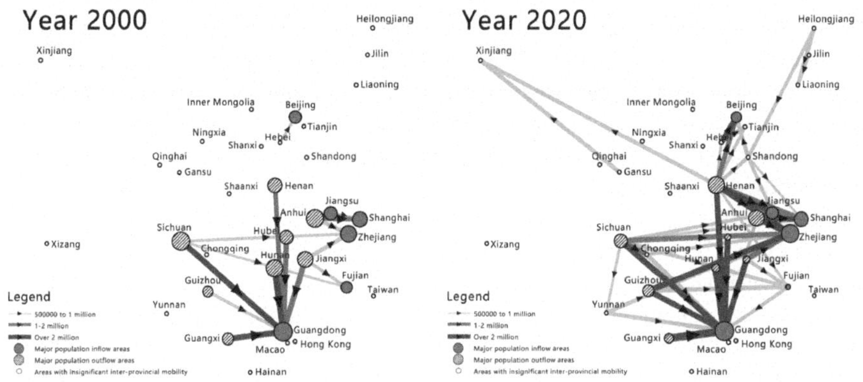

Fig. 1.2 Number of people moving across provinces, 2000, 2020. *Source* Mapping based on data from the fifth and seventh national population censuses

a significant increase in people who work and live in places other than their registered residence.

Population mobility patterns have become increasingly complex. From 2010 to 2020, the rural-to-urban migrant population grew from 143 to 249 million, with the majority being rural-to-urban migrants, primarily consisting of migrant workers. Interurban migration increased from 44 to 82 million, with an average annual growth rate of 6.42%. In terms of migration distance, as the economic development level in central and western provinces improved, the proportion of intra-provincial migrants gradually increased, rising from 61.2% in 2010 to 66.8% in 2020 among all migrants. In contrast, the proportion of inter-provincial migrants absorbed by economically developed eastern regions decreased from 82.5% in 2010 to 73.5% in 2020 (Fig. 1.2).

1.1.2 Transition from Universal Urban Growth to Increasing Differentiation

Mega and super cities have become the main hubs for population and economic concentration. By 2020, China had 105 cities with a permanent population of over one million, with a focus on mega and super cities with populations exceeding five million.[3] According to the criteria set by the State Council in 2014, cities with a permanent population of over ten million are classified as mega cities, while those with a population between five and ten million are classified as super cities. In 2020, China had 21 mega and super cities, with their permanent population increasing from 190 million in 2000 to 290 million, accounting for 20.7% of the national population.

[3] According to the "Notice on Adjusting the Standards for Urban Scale Classification" issued by the State Council in 2014, cities with a permanent population of over 10 million in the urban area are classified as megacities, while cities with a permanent population between 5 and 10 million in the urban area are classified as large cities.

Their regional GDP reached 33.4 trillion CNY, accounting for about one-third of the national total. From 2010 to 2020, the top ten cities with the largest population growth in China were all mega and super cities, accounting for 11.2% of the national population in 2020,[4] compared to 8.7% in 2010. Central and western cities became the main destinations for population inflows, strengthening the trend of population concentration in larger cities.

The emergence of giant city regions has become a significant regional phenomenon. China's five major city clusters—the Beijing-Tianjin-Hebei region, the Yangtze River Delta, the Pearl River Delta, the middle reaches of the Yangtze River, and the Chengdu-Chongqing region—had a combined population of 680 million in 2021, accounting for 48.2% of the national population, up from 30.0% in 2000. These city clusters also contributed to around 60% of the country's economic output. China's city clusters are characterized by large population size, high population density, and extensive urban sprawl. For example, the Pearl River Delta city cluster,[5] which includes nine cities such as Guangzhou and Shenzhen in Guangdong Province, covers approximately 55,000 km^2 of land area and has a population density of 9449 people per square kilometer, ranking first among global giant city regions. The Yangtze River Delta city cluster covers an area of about 212,000 km^2,[6] with a land area of approximately 29,000 km^2 and a population of around 165.08 million. The Beijing-Tianjin-Hebei city cluster encompasses the entire administrative jurisdiction of Beijing, Tianjin, and Hebei Province, covering approximately 25,000 km^2 of land area and having a population of 110.37 million.[7] In comparison, the Tokaido Metropolitan Area[8] in Japan, the Northwestern European city cluster, and the Boston-Washington Metropolitan Area in the United States have similar levels of land development intensity[9] (ranging from 10.8 to 17.3%), but their population densities of permanent residents reach 5731 and 4387 people per square kilometer,

[4] The ten fastest-growing cities in terms of population from 2010 to 2020 are Shenzhen, Chengdu, Guangzhou, Zhengzhou, Xi'an, Hangzhou, Chongqing, Changsha, Wuhan, and Foshan.

[5] The Pearl River Delta urban agglomeration includes nine cities, including Guangzhou and Shenzhen in Guangdong Province. It covers a land area of approximately 55,000 km^2, with a built-up area of about 8,000 km^2 and a total population of around 78.0 million people. The land data is sourced from the 2020 Global 30 m Land Cover Fine Classification Product. Unless otherwise specified, the construction land data for other urban agglomerations also comes from this source.

[6] The Yangtze River Delta urban agglomeration covers an area of approximately 212,000 km^2, with a built-up area of about 29,000 km^2 and a total population of around 165.1 million people.

[7] The Beijing-Tianjin-Hebei urban agglomeration includes the entire administrative jurisdictions of Beijing, Tianjin, and Hebei Province. It has a built-up area of approximately 25,000 km^2 and a population of 110.37 million people.

[8] The Tokaido urban agglomeration in Japan includes Tokyo, Kanagawa Prefecture, Saitama Prefecture, Chiba Prefecture, Osaka, Hyogo Prefecture, Nara Prefecture, Aichi Prefecture, Gifu Prefecture, Mie Prefecture, Kyoto, Shizuoka Prefecture, Shiga Prefecture, and Wakayama Prefecture. It has a built-up area of approximately 14,000 km^2 and a total population of around 71.395 million people.

[9] Land development intensity refers to the proportion of built-up area to the administrative area.

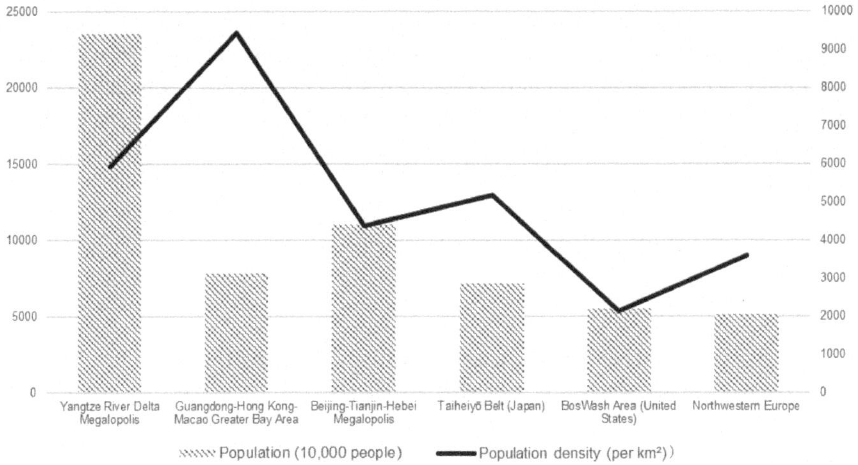

Fig. 1.3 Comparison of population size and population density in major urban agglomerations

respectively, significantly higher than the Northwestern European city cluster[10] (3608 people per square kilometer) and the BosWash Metropolitan Area[11] (2146 people per square kilometer), and comparable to the Tokyo Metropolitan Area (5175 people per square kilometer) (Fig. 1.3).

Decline begins to emerge in certain regions and rural areas. China currently has 337 cities at or above the prefecture level, with 146 cities experiencing a decrease in permanent residents and a total population decline of 37.45 million between 2010 and 2020. Among the approximately 1300 county-level cities, 174 of them have seen a decrease in population. These cities and county-level cities with population decline are primarily located in the Northeast region, border areas, and traditional agricultural regions.

1.1.3 The Belt and Road Initiative Has Raised the Development Level of Urban Areas Across the Central and Western Regions

The construction of five Central-European Freight Train Consolidation Centers represented by Xi'an, Zhengzhou, Chongqing, Chengdu, and Urumqi has enhanced the

[10] The Northwestern Europe urban agglomeration includes Greater Paris, the Rhine-Ruhr region, the Netherlands-Belgium region, and other areas. It has a built-up area of approximately 14,000 km^2 and a total population of around 51.282 million people.

[11] The BosWash Area in the United States, from Boston to Washington and including New York City, Philadelphia, and Baltimore has a built-up area of approximately 26,000 km^2 and a total population of around 54.798 million people.

level of openness and trade facilitation in these cities. The concentration of outward-oriented industries and trade enterprises has improved their functional status within the national urban system. The development of border economic cooperation zones represented by Yining and Tacheng in Xinjiang, Dongxing and Pingxiang in Guangxi, and Hekou and Ruili in Yunnan has facilitated the gradient transfer of industries from the eastern regions and enhanced the external radiation capabilities of border cities and towns. This has provided a more robust support for the development pattern of China's dual-directional openness and land-sea coordination.

1.2 Analysis of the Drivers and Policy Mechanisms Influencing China's Spatial Pattern

1.2.1 Transition from Exogenous to Multiple Dynamics in Development Drivers

Industrial upgrading has strengthened the development capabilities of cities and regions. Since China's accession to the World Trade Organization (WTO) in 2001, market openness has driven industrial expansion and technological progress. The value-added of manufacturing has grown from $476.7 billion USD in 2000 to nearly $5 trillion USD in 2021, accounting for nearly 30% of global manufacturing share and maintaining the top position worldwide for 13 consecutive years.

Innovation has become a crucial driving force for development. China has achieved international leadership in the production of advanced products in 60–80% of industrial categories through independent technology. Technologies in areas such as new energy vehicles, new materials, high-speed railways, wind power components, drones, and numerous smart and intelligent products are at the forefront globally. These high-level manufacturing industries are concentrated in the Yangtze River Delta, the Guangdong-Hong Kong-Macao Greater Bay Area, and have achieved significant development in key cities such as Beijing, Tianjin, Chongqing, Wuhan, Chengdu, Quanzhou, Qingdao, Yantai, Hefei, and Xi'an (Fig. 1.4).

Consumption is playing an increasingly important role in optimizing urban spatial patterns. In 2000, China's final household consumption expenditure was $506.2 billion USD, which reached $5.4 trillion USD in 2023, ranking as the world's second-largest consumer market, only behind the United States' $14 trillion USD. Consumption has driven urban economic development and optimized urban spatial patterns. With the increase in income levels, Chinese residents have shifted their consumption structure from basic needs to pursuing quality and branded products, undergoing an upgrading process from product consumption to service and experiential consumption. Cities are not only concentrated places for consumers but also large-scale consumer markets. Many major cities in China, known for their avant-garde, fashion, and young population, lead consumption trends, enhance development vitality, and improve quality of urban life.

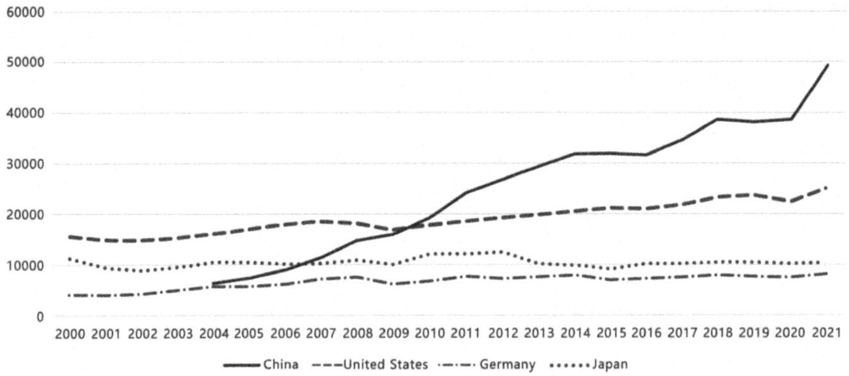

Fig. 1.4 Changes in manufacturing value added by country since 2000 (in billion USD). *Source* Statistics based on data released by the World Bank

Land finance injects strong momentum into the construction of new cities and new districts. By the end of 2018, there were as many as 3846 "new cities and new zones" constructed under various names such as development zones, high-tech zones, industrial new towns, key functional districts, and suburban new towns, with a total built-up area of 29,000 km^2 New cities and new districts have become important carriers for population and industrial agglomeration, optimizing urban spatial structures, and improving urban infrastructure and public services. The construction of new cities and new districts requires massive capital investment, and the public ownership nature of urban land in China serves as an ideal financing platform. The revenue from the transfer of state-owned land use rights increased from 50.8 billion CNY (approximately $6.1 billion USD) in 1998 to a peak of 8.7 trillion CNY (approximately $1.35 trillion USD) in 2021, providing strong financial support for the construction of large-scale new cities and new districts (Fig. 1.5).

1.2.2 Policy Mechanisms Focusing on Regional Coordination and Balanced Development

The national regional policies have become more refined. Since the beginning of the reform and opening-up, China has gradually formulated four major regional development strategies, namely, the eastern region taking the lead in modernization, the western region development, the revitalization of the northeastern old industrial base, and the rise of the central region. However, with the intensification of differentiation within these four regions, the rapid rise of mega-city regions, and the prominent contradictions between regional coordination and internal and external factors, the existing regional development strategies have struggled to adapt to the

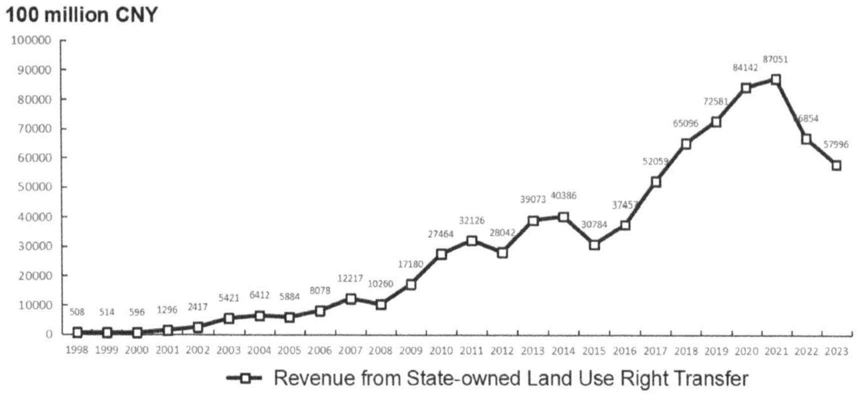

Fig. 1.5 The scale of land transfer fees for state-owned land (1998–2023). *Source* Compiled based on data from the National Bureau of Statistics

new changes in the national spatial pattern and socio-economic conditions. Consequently, on the basis of continuing the four major regional development strategies, the central government has introduced more distinct regional policies since 2014, starting with enhancing the targeted nature of regional policies. These include policies such as the coordinated development of the Beijing-Tianjin-Hebei region, the integrated development of the Yangtze River Delta, the Guangdong-Hong Kong-Macao Greater Bay Area, the Chengdu-Chongqing twin-city economic circle, and the China (Hainan) Pilot Free Trade Zone. Furthermore, in response to the increasing need for basin-wide coordination, the national government has successively issued development plans for the Yangtze River Economic Belt and the Yellow River Basin. These plans aim to optimize industrial and urban layouts based on natural geographic patterns, promote the concentration of population and economy in line with the ecological background of the basins, and have played a positive role.

The policy of functional zoning has been continuously improved. In 2010, China introduced the national strategy of functional zoning, which classifies the country's land space into regions for optimized development, key development, restricted development, and prohibited development based on development approaches. It also divides the land space into urbanization areas, major agricultural production areas, and key ecological functional areas based on development content, with the goal of promoting regional coordinated development. The areas designated for optimized and key development primarily focus on urbanization and play a key role in accommodating more industries and population while creating value. The regions under restriction and prohibition mainly involve major agricultural production areas and key ecological functional areas, where the primary responsibilities include providing agricultural products and creating more ecological products. However, these areas often bear the opportunity costs of sacrificing development in order to protect high-quality arable land and ecologically important areas, which would be unfair without appropriate compensation. Therefore, to promote and implement the functional zoning strategy, the central government has provided financial transfer payments as subsidies to counties and districts that bear the responsibility for agricultural production and ecological protection. Since 2012, the central government has allocated over 200 billion CNY annually for ecological compensation. Additionally, since 2005, the central government has implemented incentive policies for cities and counties with relatively high grain production, with the scale of incentive funds increasing from an initial 5.5 billion CNY to 46.7 billion CNY in 2020, with a cumulative total of 500 billion CNY allocated.

1.2.3 Rapid Improvement in Urban and Regional Accessibility

China's expressway network has covered all cities with a population of over 200,000, effectively supporting personnel and economic exchanges between cities. By 2022, a total of 43,000 km of high-speed railways, with speeds exceeding 250 km per hour, had been constructed, nearly covering all cities with a population of over 500,000. As of 2020, China's total mileage of expressways reached 153,000 km, ranking first in the world and 3.7 times the mileage in 2005. Among them, the density of expressways in provinces east of the Hu Huanyong Line in China reached 272 km^2 per 10,000 km^2. Currently, adjacent provincial capital cities have achieved a travel time of 1–2 h, and the central cities within urban clusters are also accessible within 1–2 h. This has promoted the formation and development of mega-city regions and enhanced the driving capacity of central cities. The status of major cities as aviation hubs has further improved, and airport distribution has become more balanced. In 2020, the top 20 airports in terms of passenger throughput accounted for 62.5% of the national total. Among the top 20 airports, the proportion of passenger throughput in airports in the central and western regions increased from 26% in 2005 to 38.5% in 2019.

1.3 Challenges in Optimizing China's Spatial Pattern

1.3.1 Addressing the Challenge of Imbalanced Regional Development

Challenges of regional economic imbalance: The southern region of China has experienced sustained economic growth, leading to a widening economic disparity between the north and south. Between 2000 and 2012, the northern region's share of the national economic output increased from 41.3% to 45.8%. However, since 2013, the economic proportion of the northern region has significantly declined, and as of 2020, the southern and northern regions accounted for 65% and 35% of the total economic output respectively. In 2022, three-quarters of the Chinese cities with a GDP exceeding one trillion CNY were located in the southern region.

China still faces imbalances in urban–rural development: In 2022, the per capita disposable income ratio between urban and rural residents in China was 2.45, significantly higher than the average level of 1.25 in most developed countries. The absolute income gap between urban and rural residents continues to widen, reaching 29,200 CNY in 2022. According to projections by the Chinese Ministry of Civil Affairs, the proportion of people aged 60 and above in China will reach 30% by 2035, which is 10 percentage points higher than the global average. This demographic shift will further contribute to population decline in rural areas and certain county towns, exacerbating the challenge of urban–rural imbalances.

1.3.2 Challenges in Addressing the Resilience of National Land Security

Ecological environment challenges: As the population and industries continue to concentrate in mega-city regions, urban and surrounding ecological green spaces are being extensively encroached upon, resulting in a widespread reduction of 3%-5% in areas such as urban water bodies and wetlands. This reduction diminishes the resilience to natural disasters and security risks. In 2020, the average economic losses caused by typhoons and floods in the Yangtze River Delta urban cluster were twice the amount in 2010. The direct economic losses accounted for 34.35% of the national total in 2020, compared to 8.36% in 2010, indicating an increasing ecological environment risk in densely populated areas.

China is a sensitive area significantly affected by climate change, facing severe threats: For instance, the rate of coastal sea-level rise in China from 1980 to 2021 was 3.4 mm/year, significantly higher than the global average during the same period (China Oceanic Administration Report, 2021). Rising sea levels intensify storm surge and have a significant impact on 130 million coastal residents, exacerbating vulnerability to tropical cyclones.

Green and low-carbon development challenges: In 2018, China ranked first in the world in terms of total coal consumption, second in total oil consumption, and third in total natural gas consumption. China's total carbon emissions were the highest globally, accounting for 27.2% of the global total in 2019. Moreover, the growth rate of carbon emissions in China is much higher than that of major world economies. From 2000 to 2018, China's total carbon emissions increased at an average annual growth rate of 11.6%, while major developed economies such as the European Union achieved negative growth. China's per capita GDP is only 1/6–1/4 of major developed economies, but its total carbon emissions are 2–3.5 times higher. According to estimates, in order to achieve the goals of peaking carbon emissions by 2030 and achieving carbon neutrality by 2060, China needs to reduce energy and resource

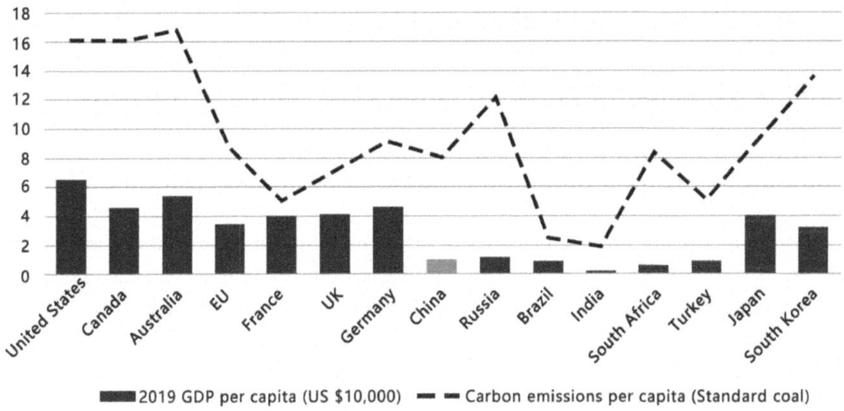

Fig. 1.6 Relationship between per capita GDP and per capita carbon emissions in major countries in 2019. *Source* Compiled based on World Bank data

consumption, as well as carbon emissions per unit of GDP by approximately 50–55% by 2035. This reduction is necessary to balance resource constraints with social and economic modernization goals. Developed countries in Europe and America achieved carbon emission peaks after reaching a per capita GDP of $30,000–$40,000 USD equivalent, by separating production and manufacturing from consumption and transferring high-carbon-consuming industries to mostly developing countries. In the next 10–15 years, China's per capita GDP is expected to approach $20,000 USD. During this period, China needs to expand production to meet domestic consumption while providing high-quality manufacturing products to the global market, resulting in significant pressure to reduce carbon emissions (Fig. 1.6).

Challenges to economic security: Influenced by the pandemic and geopolitical factors, unilateralism, protectionism, and the rise of anti-globalization sentiments have led to significant adjustments in global economic development and layout. The fragmentation of global supply chains and the rise of trade protectionism have had a sustained impact on global trade and investment activities, posing significant challenges to the security of China's industrial chains and supply chains. China's dependence on foreign sources exceeds 50% for high-end equipment, critical components and parts, key materials, and other items, exposing it to risks such as supply disruptions and sanctions. In terms of spatial distribution, China's high-end manufacturing industry accounts for 72.7% in coastal areas, and emerging industries account for 68.0%. Excessive concentration in coastal areas may bring potential risks to industrial and supply chains. Therefore, in response to the latest changes in the international situation, China has proposed accelerating the establishment of a new development pattern with domestic circulation as the mainstay and domestic and international circulations reinforcing each other.

1.4 Prospects of National Spatial Patterns

1.4.1 Establishing Future-Oriented National Spatial Patterns

In service of the millennium plan for sustainable development of the Chinese nation and centered around the Beautiful China goal for 2035, the governance of national territory is driven by "green and smart" technological innovations under the concept of ecological civilization. The aim is to build a national spatial protection and development system that coordinates regional development with national land security, aligns urban spaces with the natural environment, and matches development and construction with resource carrying capacity. This will promote the coexistence of ecological, urban, and cultural spaces.

Continuing to focus on urban clusters and metropolitan areas as core carriers, a major productive layout will be constructed with multiple support points, complementary advantages, and high-quality development to enhance economic security. The optimization of the Beijing-Tianjin-Hebei, Yangtze River Delta, and Guangdong-Hong Kong-Macao urban clusters, with a particular emphasis on upgrading the central and western regions of the Yangtze River and Chengdu-Chongqing urban clusters, will form a relatively comprehensive and multi-support industrial layout. The central and western regions of the Yangtze River will actively undertake industrial transfers from the eastern coastal areas. Efforts will be made to establish important advanced manufacturing bases and core competitive technology innovation hubs. The Chengdu-Chongqing urban cluster will develop a relatively complete regional industrial chain and supply chain system, world-class advanced manufacturing clusters, promote the development of the digital economy, and establish a financial center in the western region. Other urban clusters and metropolitan areas will serve as supplementary components, leveraging comparative advantages and cultivating distinctive industries. The current disadvantageous situation of China's strategic emerging industries being overly reliant on coastal regions will be changed by fully utilizing the supportive role of central and western hub cities such as Chongqing, Wuhan, Chengdu, Xi'an, and Zhengzhou, addressing the "weak links" in industrial and innovation chains and actively cultivating emerging industry clusters.

A national urban–rural historical and cultural heritage protection pattern will be constructed, consisting of cultural heritage corridors and cultural districts. Various cultural heritage corridors such as major rivers, ancient post roads, and border defense systems will connect cultural heritage and resources, strengthening the systematic protection of cross-regional cultural heritage corridors such as the Grand Canal, Great Wall, Silk Road, and Tea Horse Ancient Road. Based on cultural geography, national cultural regions will be defined through multidimensional factors such as dialects, ethnic groups, and architecture. These regions include the ancient capital cultural region of Beijing, the Xi'an-Luoyang-Zhengzhou-Kaifeng ancient capital cultural region, the Qilu cultural region, the Yangtze River Delta water town cultural region, the Fujian-Guangdong maritime cultural region, as well as sub-regional cultural regions along the Middle East Railway, Hexi Corridor, north and south of Tianshan Mountains, Sichuan Basin, Yunnan-Guizhou, Qinghai-Xizang Plateau, Inner Mongolia Plateau, and Xing'an Mountains. China's profound 5500-year history and cultural heritage will be protected and integrated into urban and rural areas at all levels.

Promoting comprehensive openness under the Belt and Road Initiative will continue to contribute to global cooperation and win–win outcomes. Efforts will be made to improve the transport capacity and customs efficiency of the New Eurasian Land Bridge, China-Mongolia-Russia Economic Corridor, China-Central Asia-West Asia Economic Corridor, China-Pakistan Economic Corridor, Bangladesh-China-India-Myanmar Economic Corridor, and China-Indochina Peninsula Economic Corridor, while continuing to enhance the construction of important ports along the twenty-first century Maritime Silk Road, facilitating the flow of goods and personnel. The infrastructure level, investment attraction quality, and employment supply capacity of border areas and cooperative development zones will be improved to enhance the well-being of people in all countries. Adapting to the global trend of climate warming, attention will be given to the opening conditions of the Arctic shipping route to create conditions for the opening of the Ice Silk Road. Therefore, the regional international city functions of Urumqi, Kunming, Kaxgar, and Nanning will be fully utilized, along with the role of border hubs. The level of openness in coastal cities such as Shanghai, Shenzhen, Tianjin, and Xiamen will be further strengthened, and border towns such as Kaxgar, Yadong, Yining, and Ruili will be developed to realize comprehensive and all-round cooperation under the Belt and Road Initiative.

1.4.2 Adapting to the Requirements of Urban Regeneration

An important characteristic of urbanization in its later stages is that the period of widespread urban expansion has come to an end, and cities face different futures, including continued growth, stagnation, or even shrinkage. To guide the sustainable growth of mega-cities, various approaches such as developing boundaries, blue-green spaces, and natural landscapes should be employed to achieve clustered urban development and prevent uncontrolled sprawl. For cities that have reached a relatively stable state, smart adjustments should be made with a focus on improving the urban living environment, emphasizing ecological restoration, cultural continuity, and distinctive character. For declining cities, smart contraction should be encouraged by transforming idle and abandoned old factory areas, residential areas, and storage areas into public spaces and green areas. The concentration of population for work and living should be guided to reduce service costs and promote the revitalization of central urban areas.

1.4.3 Meeting Diverse Space Requirements of the People

The further development of Internet technology has made remote work and home offices feasible, providing people with greater freedom. The flexibility of the service economy and the shortage of labor have resulted in a larger population engaged in flexible employment, allowing people to have more leisure time. With a sense of being "more free" and having "more leisure," there is a more urgent demand for aesthetically pleasing ecological products. Beautiful spaces and services that effectively meet these needs often serve as the key driving force for stimulating consumption and economic growth. China is one of the countries with the most diverse natural landscapes and cultural resources globally. By effectively harnessing the intrinsic values embedded within these resources, they can substantiate the sustainable development of the national economy while concomitantly fostering the welfare of the populace.

1.5 Summary

In the face of issues such as excessive resource consumption and environmental degradation during China's rapid urbanization process, I proposed a theoretical approach based on ecological baseline analysis to establish a coordinated national urban spatial pattern. This approach was successfully applied in the compilation of the "National Urban System Planning (2006–2020)". Building upon this foundation and considering China's geographical diversity and significant regional differences, I further refined a planning theory of "precise adaptation" between urban and natural environments, forming a methodological framework of "precise analysis - adaptive distribution - dynamic evaluation." This framework involves quantitatively analyzing the baseline conditions of regional spatial resources, developing adaptive urban spatial distribution, and dynamically evaluating and optimizing subsequent spatial developments to achieve precise, rational, and dynamic adaptation between urban, ecological, economic, and social systems.

In recent years, the theoretical approach of "precise adaptation" has been continuously improved to address the challenges of refined land governance. Frequent natural disasters caused by climate change and the concentrated outbreak of safety issues accumulated over the past 30 years of large-scale urban development have made safety a focal point of "precise adaptation". It proposes conducting regional safety risk analysis and adopting urban distribution and construction models that correspond to regional safety risks. In response to issues such as the decline of cultural characteristics due to the neglect of regional culture in urban distribution and construction, "precise adaptation" expands its scope to include culture. It suggests conducting analysis of regional cultural spatial resources and constructing urban spatial patterns and cultural spatial systems that correspond to diverse regional cultures. Looking ahead, in the context of globalization, informatization, and intensified climate change, China's urbanization and land governance will face even more formidable challenges.

The theoretical approach of "precise adaptation" needs to be continuously refined through practical application.

Chapter 2
The Origin of the Issue and Theoretical Framework of Urban Spatial Development

2.1 Questions Raised

With the advancement of economic globalization and regional integration in the twenty-first century, sustainable development has become the consensus of the international community, a new understanding generally accepted, that is, the rivalry between countries is first and foremost manifested in the competition of overall national strength with central cities at the core. And in many countries, the development of urban space has become the area where various theories and practices of urban and regional planning are experimented. Spatial planning, especially at the national level, has gradually become an important means and measure for governments to promote economic development and social progress, coordinate uneven regional development, and achieve sustainable development; it is therefore a key component of public policies.

The past decades since the Reform and Opening Up saw China's booming economy and remarkable progress in urbanization. Its urban population grew from less than 200 million to 547 million and the urbanization rate from 18% in 1978 to 41.7% in 2004; both the urban and rural landscape have undergone tremendous changes. But a host of problems cropped up because of this urbanization of unprecedented scale, such as serious waste of resources, destruction of ecological environment, uneven regional development, divorce between urban and rural areas, loss of local cultural charms, serious stratification of social structure, etc., which will hamper the rapid development of China's socioeconomic development in the next 20 years. By the end of 2004, China's per capita GDP has reached 1000 USD, marking the country's entering a crucial turning point in social and economic development, and its urbanization at this point of time was also entering an internationally recognized stage of rapid development.[1] Therefore, how to objectively judge the speed and scale of China's urbanization from the perspective of resources and environment and make

[1] Internationally, it is generally believed that a per capita GDP ranging from $1000 to $3000 marks a turning point in social development. Ray Northam, an American expert in urban geography,

K. Wang, *National Spatial Planning in China*, Urban Sustainability,
https://doi.org/10.1007/978-981-97-7729-7_2

overall arrangements for a series of weighty matters such as population distribution, industrial development, infrastructure construction, environmental protection and coordinated regional development from the spatial angle have become important considerations of the national macro-control policy.

The third plenary session of the 16th Central Committee of the Communist Party of China (CPC) in October 2003 put forward the important idea of Scientific Outlook on Development, making major adjustments to the line of thought guiding the rapid development over the past 20-odd years, putting more emphasis on the quality of development. In terms of urban development policy, it is further proposed to improve the comprehensive carrying capacity of cities and towns, and actively and steadily promote urbanization under the principles of gradual improvement, land conservation, intensive development, and sensible layout.[2] The *11th Five-Year Plan* for national economic and social development adopted by the Fourth Session of the Tenth National People's Congress in March 2006 has the chapter on "promoting coordinated regional development" specially added with such new content on spatial planning as the main functional zoning of the national land, fostering new urban clusters, and giving priority to the development of transportation industry, all of which demonstrate that spatial planning, a means of macro-control, is gradually becoming an important way for the central government to administer macro-management under the market economy. In fact, the former State Development Planning Commission (now the National Development and Reform Commission) has started the theoretical research on the reform of planning system as early as 2001, proposing to incorporate spatial planning into the *Outline of the Five-Year Plan* for national economic and social development and putting forward the concept of "national master plan" [1]. Based on its work on "Land Use Master-plan" over the years, the then Ministry of Land and Resources phased in land planning on a trial basis in some provinces and cities. In compliance with the *Urban Planning Law*, the former Ministry of Construction began the preparation for "national urban system planning" in 1994, officially started the formulation in 1999, and adjusted, reviewed, and passed the outline in 2005. The new edition of planning outline also serves as a thorough planning of national spatial resources gearing towards urban development. The separate research and formulation of overall spatial planning by different departments has incurred many problems such as overlapping functions, duplicate planning, unclear management responsibilities and so on. Therefore, how to, under the guidance of the Scientific Outlook on Development, make comprehensive judgement about the trend of urbanization from various aspects including resources, environment, population, and industry, put forward a systematic and complete national spatial development policy, and build a national spatial planning system conducive to sustainable development is not only the natural need of coordinated social and economic development of the country, but also a major theoretical and practical challenge for China's

portrayed the urbanization process as an attenuated S-shaped curve—urbanization would accelerate when the proportion of urban population reaches about 30%.

[2] See the *Communiqué* of the Fifth Plenary Session of the 16th Central Committee of the Communist Party of China.

urban and regional planning. In a broader perspective, this is also a prerequisite for comprehensively improving the construction of living environment in China.

2.1.1 *Major Problems Emerging During China's Urbanization*

The progress of China's urbanization has been increasingly faster since 1978, and by 2004, the urbanization rate increased by an average of 0.87 percentage points per year, twice that of the world in the same period. Urbanization has therefore been considered one of the two most influential events of the world in the twenty-first century.[3] Although the unprecedented urbanization has promoted the country's industrialization and alleviated the employment pressure arising from surplus rural labor, it also brought a host of problems, which, from the perspective of urbanization and urban development, are manifested mainly in the following areas:

(1) **Serious Waste of Land Resources and Extensive Mode of Urbanization**

First, with the continuous progress of industrialization, a crop of new urban areas and various development zones at all levels come into being. Due to the flaws in China's current land administration system, waste of land during urbanization is prevalent. The development zones are numerous in number and take up enormous amount of land. As of June 2004, there were 6866 development zones of various levels across the country, with an earmarked land area of 37,500 km^2, exceeding 31,500 km^2, the total area of existing urban construction land of the country in 2003. Through the clean-up and improvement efforts of the then Ministry of Land and Resources and the former Ministry of Construction in 2004, a total of 4813 development zones of various types were abolished, accounting for 70.1% of the total, and the land area earmarked for development zones was canceled by 24,900 km^2, accounting for 64.4% of the area originally planned.[4]

Second, land in the development zones is not used in an efficient way, and the benefits yielded vary greatly. Judging from the output value per unit area of the 42 state-level high-tech development zones in 2002, the industrial output value per square kilometer in the eastern region was 2.12 billion CNY, 1.05 billion CNY in the central region, and only 770 million CNY in the western region. The high-tech zone with the lowest output value managed a mere 300 million CNY per square kilometer. In general, the benefit produced per unit area of the high-tech zones in the western region is about one third of that in the eastern region (Fig. 2.1).

[3] Joseph E. Stiglitz, winner of the 2001 Nobel Prize in Economics, believes that the two major events having an impact on the world in the twenty-first century are United States' high technology and China's urbanization.

[4] According to the statistics of the former Ministry of Construction, 660 cities have a total built-up area of 25,900 km^2, and more than 18,000 towns have a built-up area of 20,300 km^2.

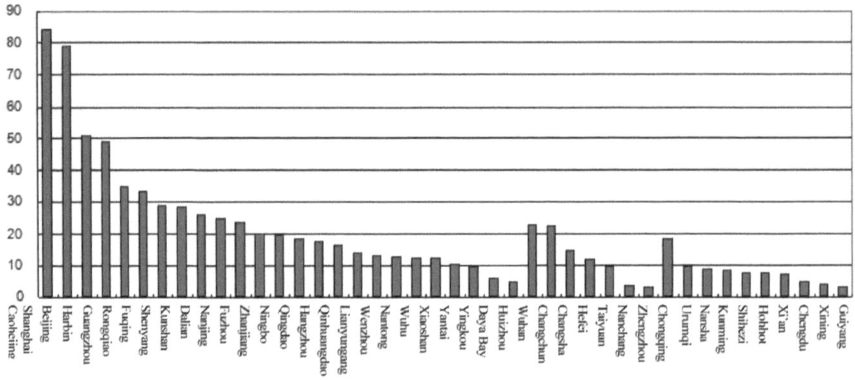

Fig. 2.1 Average industrial output value of national high-tech zones, 2002[5] (100 million CNY/km^2). *Source* Comprehensive Report on the Research of Sustainable Development Strategy during China's Urbanization. 2005. Chinese Academy of Engineering

Third, the serious problem of leaving urban land idle. Because the urban land market is in a transitional period at present, the "double-track mechanism" of land administration, development, and construction currently in use has resulted in laying idle the urban land in some cities. According to relevant survey, among the 900-odd development zones above the provincial level, the area that has been developed accounts for only 13.5% of the earmarked area. In 1999 and 2000, the former Beijing Municipal Land and Housing Administration claimed back 20 lots of earmarked land that had been laying idle, totaling 1,334,790 m^2. Besides, speculation on the real estate market is another reason that the land of some localities failed to fulfill its functions, which in turn hindered the utilization efficiency of urban land.

Fourth, the spatial organization of urban land at the regional level is neither sensible nor intensive. Due to the lack of regional planning and the fact that local land is subject to administrative divisions and market barriers, the allocation of regional land resources failed to meet the development needs of rapid industrialization and urbanization. Same projects and similar infrastructure construction are repeated, which not only led to over-exploitation of land resources in certain localities, but also incurred the waste of land resources in general.

(2) **Serious Environmental Pollution and General Decline in the Functions of Regional Ecological and Environmental Service**[6]

First, air pollution is so serious as to threaten the health of people. According to monitoring, 58.3% of China's 340 cities had an air quality inferior to the national secondary ambient air quality standard (standard for residential areas) in 2003. Among the 113

[5] 42 out of the total 46 of the country. The cities of Tianjin, Weihai, Nanning, etc., are missing due to unavailable information.

[6] Source: Comprehensive Report on the Research of Sustainable Development Strategy during China's Urbanization. 2005. Chinese Academy of Engineering.

key cities for air pollution prevention and control, 66.2% had an air quality that is not suitable for human habitation, and 73.9% of urban residents lived in an atmospheric environment that are not suitable for human habitation. Air pollution has led to the increasingly serious atmospheric phenomenon of haze, which was prevalent throughout North China, Central Plains, South China, and East China in recent years. Regional haze situation is especially frequent in built-up urban areas, take Guangzhou in the Pearl River Delta for example—in 2002 and 2003, the city had the haze situation for periods as long as 7 and 20 consecutive days, respectively, and it was even more frequent during September to November 2004. Atmospheric haze has become one of the major climatic and meteorological disasters in the urban clusters in Pearl River Delta. Regional haze of such a degree is few and far between in other countries and regions in the world.[7]

Second, the cities are facing serious water crises given water resources are scarce and its utilization efficiency low. China has a per capita water resource of 2,304 m^3, which is one fourth of the world's level, and the extremely uneven temporal and spatial distribution of the resource only exacerbated the problem. According to statistics, more than 300 of 661 Chinese cities are short of water, and 114 are seriously so. Worse still, due to over-exploitation of groundwater in some areas, large groundwater funnel areas are forming; more than 50 "funnels" have been formed in 10 provinces, municipalities, and autonomous regions in the north, with a total area of 30,000 km^2. In the Yangtze River Delta, over-exploitation of groundwater has led to land subsidence of 8000 km^2, incurring economic losses of about 315 billion CNY. To be specific, three regional subsidence centers have formed in the urban area of Shanghai, Suzhou-Wuxi-Changzhou region of Jiangsu and Hangjia Lake of Zhejiang, respectively, and they threaten to develop into a continuity; the funnel center with the direst situation (Luoshe Town, Wuxi) has a groundwater buried depth of 84 m.[8]

Third, cities are encircled by garbage as the solid waste is rapidly increasing. According to the monitoring results of 46 key cities for air pollution control by the State Environmental Protection Administration (SEPA) in 2002, the rate of harmless treatment was less than 15%. The cumulative stock of urban waste reached 6 billion tons, covering an area of more than 200 km^2, and it keeps growing at an average annual rate of 4.8% in recent years, which rendered nearly two thirds of the country's cities to be surrounded by garbage. Every year, 79 million tons of domestic waste are landfilled in the simplest way or just dumped to pile up in the open on the outskirts of cities and along rivers, causing a series of problems such as water pollution, decline in water quality, soil pollution and the spread of infectious disease, which jeopardized the sustainable development of urban society and economy.

Fourth, serious pollution of the offshore waters is damaging the environmental quality of coastal cities. Major eutrophication has occurred in China's coastal waters,

---

[7] Source: Comprehensive Report on the Research of Sustainable Development Strategy during China's Urbanization. 2005. Chinese Academy of Engineering.

[8] See "Prevent the Good Land from Turning into Waters - Reflection on Land Subsidence in the Yangtze River Delta", *Geological Exploration Guide*, January 11, 2005.

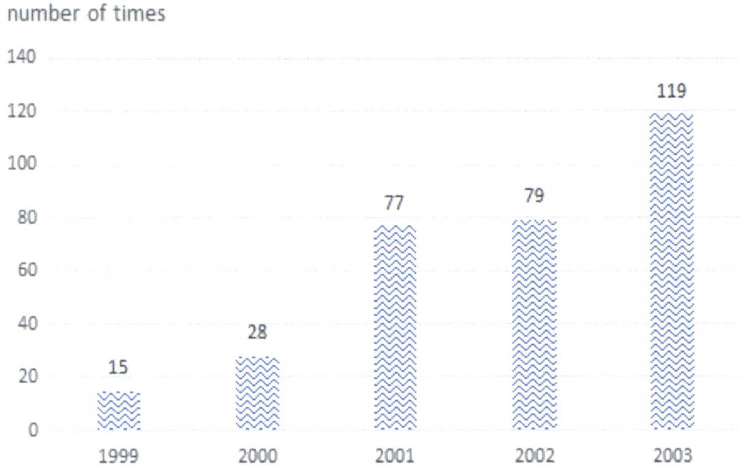

Fig. 2.2 Occurrence of red tides in China, 1999–2003

and the frequency of red tides has been increasing (see Fig. 1.2). In 2003, the cumulative area of red tides reached 14,550 km^2, and it was characterized by long periods, concentrated high-incidence period, and extended duration. Red tides occur in all the 12 months of the year and last up to 35 days (Fig. 2.2).

(3) **Increasingly Widening Gap between the East, the Central and the West, and Seriously Imbalanced Regional Development**

First, new imbalances in China's regional pattern occur due to the rapid advancement of urbanization. The eastern coastal region, thanks to its geological location and advantages in terms of capital and favorable policy, achieved rapid development with its scale of urban population and construction land and number of cities and towns increasing significantly. Taking the Yangtze River Delta for instance, in 2004, the total economic volume of the two provinces of Jiangsu and Zhejiang and the municipality of Shanghai reached 3.4 trillion CNY, accounting for 24.99% of the national economy.[9] The central and western regions, however, were relatively slow in urban development due to weak economic foundation and relatively backward policies, and a new round of migration and the so-called "quasi-urbanization" is taking place as rural surplus labor are relocating from west to east and north to south on a large scale across various regions. According to the statistics of the "Fifth Census" (2000),[10] of the three major regions of the east, the central and the west, the eastern region enjoyed the highest urbanization level of 44.60%, followed by the central region with 33.50%, and the lowest in the west at 27.66%, indicating

[9] See the *2005 Statistical Yearbook* of the whole country and the cities and provinces of Shanghai, Jiangsu, Zhejiang and so on.

[10] In the Fifth National Population Census (2000), urban population refers to those who live in the cities for more than half a year, which was relatively objective, but the criterion was not adopted in the censuses of subsequent years.

a difference of nearly 17 percentage points between the east and west. The gaps were even greater from province to province. The huge gap between regions and the serious lag in urban development in the central and western regions have been harming the healthy social and economic development of the country.

Second, although the central government made major decisions on regional development in succession since 1980s such as the trailblazing development of the eastern region, the large-scale development of the western region, the revitalization of the old industrial bases in the northeast, and the rise of the central region, aiming to redress the imbalance of development between regions, the lack of spatial policies complementary to those development strategies made it impossible to reflect the natural requirements of coordinated regional development in practice. Besides, given the improvement of the market economy system over time and gradual deepening of economic globalization of today, this kind of "improvised" regional policies over the past 20 years needs to be re-sorted from the perspective of resources and environment and based on the analysis of the economic development level of different regions. Meanwhile, regional policy zoning needs to be redefined considering the core status of central cities in regional economic development to orchestrate regional development cross the country (Fig. 2.3).

(4) **Widening Gap Between Urban and Rural Areas and Prominent Problem of Urban–Rural Dual Structure**

Due to the dual structure divorcing urban and rural areas, the gap between urban and rural areas across China keeps widening over the past 20 years. The income

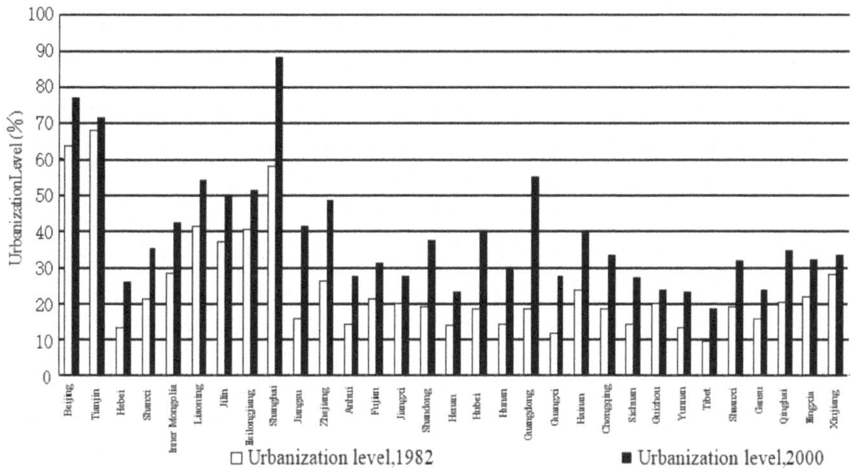

Fig. 2.3 Urbanization level of 31 provinces, autonomous regions, and municipalities, 2000. *Source* Comprehensive Report on the Research of Sustainable Development Strategy during China's Urbanization, 2005. Chinese Academy of Engineering

gap widened from 2.33:1 in 1992 to 3.2:1 in 2004.[11] The gap in the quality of living environment is even greater, which is reflected in all the following aspects: education, culture, health, social security, science and technology, and quality of life. Most rural residents have no access to the basic public health and medical care. The rural population, which accounts for 70% of the total, only consumes 20% of the medical products and services. In terms of the distribution of educational resources, the average number of years of education received by the rural population aged 15 and above is less than seven. There are 85 million illiterate and semi-literate people nationwide, more than three fourths of which were in the western rural areas, areas inhabited by ethnic minorities and national-level poverty-stricken counties.

Meanwhile, the large number of migrants from rural areas have suffered unfair treatment. Although some of them have managed to change occupation and settle down in the cities, they are only regarded as "quasi-urbanized population" without any real change in their identity,[12] thus forming a "dual structure" within the cities. The "threshold" to obtain citizens' rights and urban social security is so high that it is difficult for them to obtain equal pay for equal work, let alone enjoy such securities and rights provided by the cities as medical care, pension, unemployment insurance, education, affordable housing and so on. Unable to fully fit in urban society, they are the perpetual outsiders in cities. Besides, their low education level and poor labor skills made it impossible for them to adapt to the change from peasants to city-dwellers and meet the requirements of urban modernization. According to the data of the Fifth National Population Census of the 144 million floating population, 61.1% have an education level of junior high school or below, rendering them unfit to serve as the labor reserves of urban enterprises for technological progress and industrial structure upgrading. If a marginalized social group continue to exist alongside the mainstream group, serious social conflict is bound to happen, and situations like "over-urbanization" in some Latin American and African countries are sure to appear.

(5) Spatial Disorder and Polarization of Central Cities Co-exist with the Decline of Small Towns

First, coastal mega-urban regions are increasingly polarized and overdeveloped. In the past 20 years, a large influx of rural surplus labor was drawn to the eastern coastal areas by its rapid economic development and exponential urban growth, resulting in the formation of such densely populated urban regions as the Yangtze River Delta, the Pearl River Delta, and the Bohai Rim with Beijing-Tianjin-Hebei as the core. Urban development is increasingly trending towards regional clustering. However, the high industrial density and population density in these regions has led to serious damage to ecological environment, as evidenced by the fact that the three regions

[11] See the speech of Gao Hongbin, Deputy Director of the State Council Leading Group Office of Poverty Alleviation and Development, at the Policy Seminar on "Pro-poor Economic Growth" of East Asia.

[12] According to the calculation of China's Fifth National Population Census (2000), urban population refers to those who live/work in the cities for more than half a year, but most of them are floating populations, so they are called "quasi-urbanized population".

mentioned above have become the most polluted offshore waters in China and the areas worst damaged by acid rain.

Second, there is a lack of integration and coordination in the cities of the regions as each of them tend to develop by itself and compete. As a result, no regional synergy is formed to help them develop into mega-urban regions with international competitiveness, which hinders the country in its international competition. Taking the Yangtze River Delta for example, in 2003, the similarity coefficient of industrial structure of Shanghai and Jiangsu was 0.82, that of Shanghai and Zhejiang was 0.76, and it was as high as 0.97 of Zhejiang and Jiangsu.[13] Apart from this, duplication of infrastructure is also a prevalent problem. Again, taking the Yangtze River Delta for instance: Shanghai's financially ambitious enterprise to build world-class international shipping center (Large and Small Yangshan Port) failed both to coordinate with Ningbo Port of Zhejiang, which has excellent conditions for port construction, and to take the construction and development of the port group in the entire region into consideration. And to strengthen economic ties with Shanghai, Ningbo, Hangzhou, and Shaoxing, three cities of Zhejiang Province, all plan to build a cross-sea bridge in Hangzhou Bay, hence the chaotic situation of three bridges in northern Zhejiang connecting with Shanghai; the latter was not at all enthusiastic about the site selection for the bridge. In regional management, although the mayor joint meeting mechanism of the 15 cities in the Yangtze River Delta has long been established, the mechanism for coordinating the overall interests of the region in terms of industrial development, infrastructure construction, and environmental protection, etc., has yet to be established, and nothing whatsoever has been planned for the organizational setup, implementation planning and operational means of regional coordination.

Third, the general backwardness of small towns has been neglected, resulting in its weak foundation of urban structure. In the past decade, the country's investment in industry and fixed asset were mainly in large and medium-sized cities, which incurred such problems in small towns as poor service functions and degradation of self-development capabilities. The general backward economy and lack of specialty industry of importance in small towns also brought the following problems: low degree of population agglomeration, small in scale, lagging infrastructure construction, and inability to attract people from the vast rural areas and industries and to provide effective space for the country's new road to industrialization. Although the number of towns has increased rapidly since 1978, reaching nearly 20,000 up to now, the average size of towns is only 1221. 1 households and a population of 4518.6 according to the data of China's first agricultural census. Such small scale meant small space for development and little influence over its surrounding area, and low benefits for resource aggregation, rendering them ineffective space for China's urbanization effort.

(6) **Extensive Mode of Urban Economic Growth and Insufficient Driver for Innovation**

---

[13] See "Integration of the Yangtze River Delta Calls for A Breakthrough", *Wenhui Daily*, February 24, 2006.

First, given its high foreign trade dependence and poor capacity for independent innovation, China is in the middle section of the global industrial supply chain, making it difficult to earn high economic benefits. Although it is generally accepted that China is the world's "manufacturing power", some experts believe that[14] considering its current situation, it can only be called a "processing power". According to relevant statistics, China is in the forefront in manufacturing equipment industry in terms of the number of power generation equipment, machine tools and automotive products, but it still depends on imports for 85% of IC manufacturing equipment, 70% + of high-end CNC machine tools, and 100% of optoelectronic manufacturing equipment. In the information industry, precision instruments, biotechnology, etc., China has capacity for only one packaging and assembly process, which means about 90% of the industrial value is generated abroad. This kind of production method and structure makes it difficult to maintain long-term healthy and rapid development.

Second, the consumption of resources for production ranks among the highest in the world. At present, China's total energy consumption is about one third of the United States, ranking second in the world and accounting for 1/10 of the world's total energy consumption. In 2003, China's consumption of crude oil, raw coal and electricity was about 7.4%, 31% and 13% of the world's total, respectively.[15] Recent years have seen frequent power shortage and oil shortage, and the "coal-power dispute" further highlighted the energy bottleneck in China's economic development. A calculation based on the current growth rate of energy consumption forecasts severe supply–demand imbalance in future. It is predicted that in 2020, China's demand for primary energy is as follows: about 2.5–3.3 billion tons of standard coal, including 2.1–2.9 billion tons of coal, 450–610 million tons of oil, 145.3–165.4 billion cubic meters of natural gas, and 860–950 million kwh of installed capacity of generators, far exceeding China's current energy development capacity and proven reserves of energy minerals.[16]

(7) **Regional Cultural Resources Damaged and Social Development Lagged Behind**

First, with the rapid advancement of urbanization, destruction events in certain long-standing cities, historical areas, cultural relics, scenic spots, natural heritage, intangible heritage, etc. Frequently happened, and some regional historical and cultural resources have also suffered major damage with the implementation of a series of major construction projects. For example, a number of cultural relics along the Three Gorges Dam Project and the South-to-North Water Diversion Project have been damaged to varying degrees.

[14] Opinion of Professor Yang Shuzi of Huazhong University of Science and Technology at the Science and Technology Forum of the 7th Shanghai International Industry Fair and the 3rd Shanghai International Conference on Vibration Engineering in 2005. *CBN Daily*, November 23, 2005.

[15] See Major Issues Concerning China's Mid-to-Long-Term Development 2006-2020, Wang Mengkui. China Development Press. 2005.

[16] See Major Issues Concerning China's Mid-to-Long-Term Development 2006-2020, Wang Mengkui. China Development Press. 2005.

Second, the loss of regional cultural characteristics. Not only did "a thousand cities share the same appearance", but many regional natural and cultural landscapes are increasingly losing their characteristics. Some areas with historical and cultural traditions are also keen on modern cultural attractions such as amusement facilities of the west, neglecting its own cultural heritage. As a result, it becomes increasingly difficult to distinguish one renown city or place from another.

Third, some regions are keen on building cultural landmark facilities such as grand theaters and convention and exhibition centers at the expense of investment in basic education and grassroots cultural facilities. On one hand, large theaters and convention and exhibition centers in central cities easily cost hundreds of millions or even billions of Chinese Yuan; on the other, such public welfare cultural facilities that have a direct bearing on the life of urban and rural residents as community-level cultural centers, libraries, and sports facilities are scarce. For example, in terms of educational facilities, the construction of "college towns"[17] is the rage, while basic education facilities, especially primary and secondary schools in villages and towns, have been neglected. As a result, the large number of rural surplus labors can only get jobs that require the minimum level of skills because they cannot meet the requirements of urban industrial upgrading.

(8) **Various Special Plans and Regional Plans Continue to Emerge while Comprehensive Planning is Urgently in Need**

First, various state-level special plans were formulated even though no analysis has been done on the spatial distribution of cities and towns across the country. The Ministry of Transport has organized to have the following formulated: 2020 national road network planning, national highway network planning and port layout planning; the former Ministry of Railways has organized the planning of the 2020 national railway network and passenger dedicated line network; the Civil Aviation Administration has formulated the 2020 national airport layout planning; the then Ministry of Land and Resources is preparing the second round of the overall national land use planning; and the State Environmental Protection Administration is preparing a number of special plans including national ecological function zoning. Although these planning are related to the country's medium- and long-term plans for national economic and social development, the organization is not sensible due to the lack of research on cities—one of the major objects they serve, resulting in such problems as sluggish development or over-investment. As the most important macro strategy for socio-economic development at the national level, the *Eleventh Five-Year Plan* takes thorough consideration of such issues as coordinated regional development and national urban axis, but the comprehensive national spatial plan leaves much room for improvement. And the national spatial development plan that can be called mid-to-long term is weaker in its existing foundation. Although a new round of land planning research is initiated by the then Ministry of Land and Resources, and the "Plan of

[17] IN recent years, more than 60 college towns (cities) have been built or are under construction across Chinese cities. According to the survey, 30 of them cover a total area of about 435 km^2, with an average area of 14.5 km^2 per park.

Beijing-Tianjin-Hebei Integrated Development" and the "Plan of the Yangtze River Delta Integrated Development" were formulated by the National Development and Reform Commission, there is still no systematic integration of spatial development planning at the national level.

Second, 27 provinces and autonomous regions have respectively prepared provincial urban system plans with urban development as the core and taking into consideration socioeconomic development, infrastructure, ecological environment and so on according to the requirements of the *Urban Planning Law* and considering the needs of their own development, which has played a positive role in promoting the coordinated development of industries and towns within the provinces. For example, the planning of the coastal provinces where the three metropolitan interlocking regions are located lacks connection and coordination with their neighboring provinces and regions in terms of infrastructure connection and resource protection and utilization, and the characteristics of cohesion are distinct; it also failed to fully consider the role of the three regions in driving the economic development of the central and western regions. Making full use of the favorable policy of "the rise of the central region", all provinces in the central region are actively building their own urban clusters and central cities, putting forward various concepts including the Urban Cluster in Central Plains, Wuhan Metropolitan Coordinating Region, and ChangJiu Industrial Corridor, respectively, resulting in a self-centered and mutually divided urban spatial form; a regional central city which is key to the rise of the region is yet to be found. The southwest region does not have conditions for centralized development due to its terrain of fragmented land. Yet, it is also planning for large-scale urban clusters without a thorough consideration of the right development mode under its special resource conditions and its economic and transport connection with the developed regions in the East and the South. The northwest region, despite its being sparsely populated with small-scale towns and large spacing between towns, is also using the axis/belt principle to plan its urban system, which is far-fetched. Besides, the attention paid to the urban development of border areas is insufficient, and there is a lack of understanding of the significance of strengthening border city construction to the overall security and energy security of the country. The three northeastern provinces of Heilongjiang, Jilin, and Liaoning attached too much importance to the integrity of its own spatial structure in the province, each putting forward an urban axis for development, and ignoring the construction of the major development axis—from Harbin to Dalian—of the entire region. It is thus difficult to achieve the purpose of improving the status and role of the northeast region in the development of northeast Asia (Fig. 2.4).

Third, in recent years, some cities and regions have been actively exploring in regional planning in its various forms, typical examples include the regional plans of Pearl River Delta Urban Cluster, Suzhou-Wuxi-Changzhou Metropolitan Region, Hangzhou Bay Urban Cluster, Harbin Metropolitan Region, Wuhan Metropolitan Region, etc. Despite their contribution to the development of regional planning, this series of research has many problems such as lack of the guidance of a master plan, inaccurate positioning, and insufficient coordination.

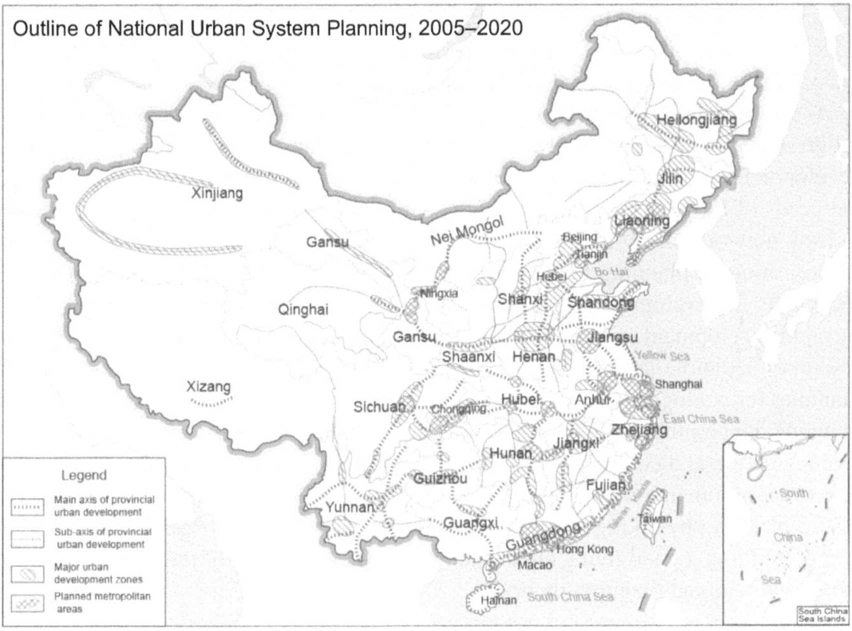

Fig. 2.4 Map of 27 provincial-level urban system planning. *Source* Outline of National Urban System Planning, 2005–2020

2.1.2 The Necessity of the Study

Under the above circumstances, China is facing unprecedented challenges in such aspects as urbanization, healthy urban development, and urban and regional planning. First, the rapid economic growth must be maintained in the next 20 years. But the development model over the past two decades cannot continue given the current resources and environmental conditions. So, major adjustments need to be made to the line of thinking concerning urbanization and urban development to arrive at a development path conducive to economic growth, more intensive use of resources, environmental friendliness, and high quality of human settlements. Second, the sharp change of urban spatial structure made it a natural for us to sort out urban spatial structure to promote healthy urbanization. On one hand, by complying with the laws of globalization and market economy, we need to accelerate the pace of development of some cities and regions and improve national competitiveness. On the other, we need to give effective play to the government's role of macro-control, preventing certain regions from being marginalized and achieving coordinated regional development. Therefore, research in urban spatial planning at the national level is indispensable. Third, various planning practices were born with the rapid urbanization of China in the past 20 years, such as land resources planning, provincial urban system planning, regional planning, urban cluster planning, metropolitan coordinating region planning, development strategy of urban space, etc. To take stock of these rich practices to innovate urban and regional planning theory with Chinese characteristics is undoubtedly of great practical significance.

As far as planning is concerned, the rapid economic development and urbanization progress of China have brought about two kinds of planning at the national level. First, development-oriented plans conducive to economic growth keeps appearing to meet the needs of rapid development, such as national industrial development planning, railway network planning, highway network planning, port planning, etc. Second, control-oriented planning for alleviating the pressure of development and ensuring the quality of development too keeps coming up to fix such problems brought about by rapid development as deterioration of ecological environment and shortage of land resources, example in this regard including the overall national land use planning, planning for ecological environmental and water resources preservation, etc. These planning concerning different aspects and subjects essentially serve as comprehensive regulation and control of national spatial resources since they all center around industrialization and urbanization. Therefore, putting forward a systematic idea for macro planning by taking thorough consideration of all the problems and with the aim of building a good living environment is the natural need of the work related to urban and regional planning theory.

2.2 Related Concepts and Theories

2.2.1 Related Concepts

Urban Space

Urban space refers to the spatial environment involved in urban development with the city as the core. It mainly includes the natural and ecological environment, transportation facilities, municipal utilities, and human resources in the vicinity of the city. In terms of geographical scope, it is larger than the built-up area and smaller than the administrative district. The concept is used in this paper mainly to distinguish it from cities in the sense of administrative region and central cities in the sense of the built-up areas, because the former lacks the meaning of physical space, and the latter only reflects the current state of development. The research on urban spatial development at the regional level used to focus on the urban system, more specifically, its functional structure, hierarchical structure, and scale structure. The national urban spatial development studied in this paper has more to do with land resources analysis, socioeconomic background research, dynamic trend of urban spatial structure and regulatory mechanism of spatial development.

Land Resources

National land resources refer to the territorial space under the jurisdiction of a sovereign state, including territory, airspace, territorial sea, and the exclusive economic zone in accordance with the *Convention on the Law of the Sea* [2, 3]. The national land in this paper mainly refers to the place for people's survival, living and production activities, including natural and human resources. So, land resources are an apt description. In a broad sense, land resources include all natural resources, social resources (e.g., human resources, social and cultural resources, etc.) and economic resources within the scope of a country's sovereignty, while in the narrow sense, it refers only to natural resources. "Land resources" in the context of "land resources planning" mainly refers to its narrow sense, so does the "land resources" in this paper in most cases, except a handful of exceptions.

Land Resources Planning

There are many interpretations of the meaning of land resources planning. According to the entry "Regional Planning", Vol. Architectural Garden, "Urban Planning", *Encyclopedia of China*, there is no essential difference between land resources planning and regional planning: apart from the difference in the size of geographical scope, both refer to the comprehensive layout planning of various construction within a certain geographical space. Therefore, under certain conditions, the two terms can be used interchangeably. According to the concept confirmed by the then Ministry of Land and Resources, land resources planning refers to the comprehensive strategic deployment of the development, utilization, improvement and protection of land resources, and also the comprehensive spatial layout of major activities of land construction, which requires the following: coordination of the relationship between resources, economy, population and environment, sound adjustment and layout of industrial structure, planning of urban system and the allocation of major infrastructure networks. The objective is to achieve sustainable socioeconomic development while ensuring harmonious coexistence of man and nature. In the systematic study of the then Ministry of Land and Resources, the spatial planning, regional planning, and development planning of some Western countries are equivalent to land resources planning.

Urban System

According to the *Standard for Basic Terminology of Urban Planning* (GB/T50280-98), urban system refers to an organic whole with the central city as the core and composed of a series of towns of different levels, scales, and functions and closely related to each other within a certain geographical range. Some scholars believe that the concept of urban system contains multiple layers of meaning: first, its research object is the urban group in a relatively complete area, and different regions have different urban systems. Second, its core is the central city: a town system with modern significance is not possible without a central city with certain economic and social influence. Third, the essential feature of the urban system is interconnection: cities and towns form vertical and horizontal connections through different locations, levels, scales, and functions, thus forming an organic whole. The reason why the urban system is called a system lies in its integrity, hierarchical levels, and dynamics [4].

Urban System Planning

According to the *Standard for Basic Terminology of Urban Planning* (GB/T50280-98), urban system planning refers to the distribution and development planning of cities and towns with different population sizes and functional division of labor within a certain geographical range based on the rational layout of regional productivity and the division of urban functions. Some scholars believe that from the regional level, urban system planning plays the important role of connecting regional land resources planning and urban master planning, therefore possessing dual nature—an integral part of both urban planning and regional land resources planning [4]. The study on "The Relationship between Regional Planning and Urban System Planning" organized by the Ministry of Housing and Urban–Rural Development holds that the above definition has the distinct flavor of the planned economy, and suggests that it be changed to: Being an important form of regional planning in China, urban system planning refers to planning that aims at the coordinated development of the towns and cities in the region and urban and rural areas, and involves the determination of strategies of regional urbanization and urban development, sound arrangement for regional urban organization, coordination of the temporal and spatial relationship between urban development and industrial allocation, and overall coordination and comprehensive arrangement of the following matters: utilization of regional land and various resources, allocation of infrastructure and social utilities, environmental protection, etc.

Spatial Planning

Spatial planning is a new interpretation of urban and regional planning in recent years. In the *European Spatial Development Perspective (ESDP)* published by the European Union in 1999, the term "spatial planning" was used to replace urban and regional planning. Although still based on physical spatial planning, it acquired some new meaning, that is greater importance is attached to the following aspects: promoting regional economic development through planning, guarantee of social benefits (such as education) and cultural diversity, ecological environment, and sustainable development, and in the means of planning more emphasis is placed on financial strength and regional infrastructure construction. More importance was attached to the concept of spatial planning in the new era. On one hand, it played the role of unifying and standardizing different planning systems of European countries, and on the other, it signaled the return to or clarified physical spatial planning. The Netherlands and Germany were among the first to have put forward the theory of spatial planning and carried out relevant practices in this regard. Following the publication of the *European Spatial Development Perspective*, the European Commission carried out several transnational and national spatial planning practices in different regions.

National Spatial Planning

National spatial planning refers to the overall comprehensive arrangement and strategic deployment for continuous coordination of the factors of economy, society, resources, and environment nationwide. It is planning at the macro-scale level, which is usually organized and prepared by the central government or other high-level institutions. A long-term basic plan of the national spatial development framework for more than 10 years, it also serves as the planning of physical space that involves comprehensive allocation of major economic activities and resources on a broad scale. The Netherlands has carried out five national spatial planning since the 1960s and its central government used it as an important means for spatial resource management, coordinated regional development and urban function adjustment. This paper borrows this concept to illustrate the spatial planning system with urban space as the core.

Region

Region is a broad concept, and its definition varies depending on the specific discipline and research object. In political science, region refers to the administrative unit of state management. In sociology, it refers to human settlements with the same language, belief system, and ethnic characteristics. In economics, it is seen as economic and social complexes with specific regional characteristics that shaped by human economic activities. In geography, it is defined as the territorial unit of the Earth's surface, and the Earth is believed to be made up of an infinite number of regions. According to Edgar Hoover, expert in regional economics, "a region is a continuum of areas that is considered useful for describing, analyzing, managing, planning, or formulating policy[18]". A more comprehensive and essential definition of "region" was proposed by the American geographer D. Whittlesey. He believes that region is "a device for selecting a studying area grouping of the complex phenomena found on the earth"; any segment or portion of the earth's surface is a region if it is homogeneous in terms of such an aerial grouping. This paper uses the concept of region to illustrate the regional characteristics of urbanization and the regional space of specific functions.

2.2.2 *Basic Theories*

The theories discussed herein mainly involves the science of human settlements, globalization theory, spatial planning theory, spatial structure theory and government intervention theory.

(1) **Science of Human Settlements**

In the final analysis, urban spatial development is about the construction of human settlements. According to the science of human settlements proposed by Mr. Wu Liangyong, human settlements refer to the human-centered living environment that covers both man and other living beings and people's living systems. Any human settlement is founded on the following five aspects: ecology, economy, technology,

---

[18] Quote from *Regional Analysis and Planning*, by Cui Gonghao. Beijing: Higher Education Press, 1999.1.

society and culture and art. This theory puts forward a new research methodology for the development of urban space at different levels. Borrowing the theory of systems science, he puts forward the idea that the human settlement is a complex giant system, and a coherent and comprehensive research method is needed; he believes that the study of cities and regions should be regarded as a science about the whole and wholeness.

As far as China's urban development is concerned, this theory holds that it is necessary to start with the national conditions, grasp the imbalance of its urbanization progress, and explore different paths of urban development. For the urban system, certain patterns can be identified in the evolution of urban forms. For instance, mega cities and large cities tend to shift from centralized and monocentric structures to dispersed and regional structures. In economically developed and densely populated areas, rural towns tend to become relatively concentrated with the spread of industry. For the entire urban system, "large, medium, and small cities need to develop in a coordinated manner to form a sound urban system, and a comprehensive network between urban and rural areas and between regions need to be established gradually to promote coordinated economic, social and cultural development of urban and rural areas" [5].

(2) **Globalization Theory**

Since the 1980s, the world economy has become increasingly integrated and in this context, it was re-organized: the financial structure becomes more centralized, and financial capitals, having exceeded all spatial and location limits, are playing an important role in and having great impact on the allocation of global resources; the status of transnational corporations in international economic activities and national power has been enhanced—the division of labor in global industries has been realized to a large extent through transnational corporations; the capitals, having stronger motive to move across national borders, are acquiring more distinct regional characteristics, and the cooperative relationship between the regional and global markets has been strengthened. The flowing of capitals has brought about the global relocation of productive space, i.e., the emergence of the so-called "space of flows," which has been manifested mainly by foreign direct investment (FDI) for nearly 20 years. Throughout the world, foreign investment is playing a role at all levels of space and featuring prominently in economic development policies of various countries, it serves as great driver for local economic development. The theory of globalization has become the fundamental theory for analyzing the evolution of regional spatial structure in developed regions under the market environment, and it also plays an important role in exploring the imbalance of regional development.

Globalization Theory holds that under globalization, the functions and status of cities on a global scale are constantly adjusting and changing, and gradually forming a new organizational structure. Friedman and Sassen proposed a new hierarchical structure of global cities from the functional properties of cities. Hall proposed that in the era of globalization, with the dispersion of production across the world, service activities are increasingly separated from the location of material production, and high-end service activities worldwide are increasingly concentrated in a few trading cities, and the management of global financial capital and corresponding services are especially so as they are increasingly centered in a handful of countries and cities. In addition, Hall and other experts also pointed out that a new urban phenomenon is taking shape amid globalization, that is the Mega-City Region, which has the characteristics of global production and service, and is playing an indispensable role in the division of labor of global production. Gradually, a spatial area composed of dozens or even hundreds of towns that serves multiple functions and boasts both high-end and low-end services is forming.

(3) **Spatial Planning Theory**

In recent years, the planning theory and practice of developed Western countries are attaching more importance to the integrity and coordination of spatial development. Returning to the planning system centering on physical space planning, they gave more value to economic, social, and environmental goals based on the original planning and named this kind of planning with integration and coordination functions and of strategic importance "Spatial Planning." This concept, initiated by the EU, has gradually become the collective term for the planning systems at various regional levels in many developed countries.

National spatial planning refers to the overall comprehensive arrangement and strategic deployment for continuous coordination of the factors of economy, society, resources, and environment nationwide. A long-term basic plan of the national spatial development framework for more than 10 years, it also serves as the planning of physical space that involves comprehensive allocation of major economic activities and resources on a broad scale. For most countries, national spatial planning is an indispensable means for the country to improve the market economy system, enhance competitiveness, and administer macro-control; it is a means for the central government to prevent and correct market failures and make government intervention under a completely free economic system. Politically, national spatial planning plays the role of embodying democracy, it is the expression of democratic ideology and popular participation. In terms of administrative system, national spatial planning is the general principle coordinating the interests of various departments of the state and between the central government and local governments, that is it has the role of administrative coordination. Spatial planning is gradually becoming an important type of planning for some countries to cope with international competition, achieve sustainable development and social progress in the era of globalization.

(4) **Government Intervention Theory**

Economic theory and the theory of political science have made some essential analyses of the intervention in space. For the relationship between the pioneers and latecomers of development, the theory of circular and cumulative causation holds that a "geographical dual economy" exists in the economy of underdeveloped countries, and that government intervention is a necessary means to promote the coordinated development of regional economy. When certain regions have accumulated development advantages, the government should adopt the strategy of uneven development and give priority to the regions with stronger growth potential, to gain higher investment efficiency and faster growth rates, and then drive the development of other regions through diffusion effects. Meanwhile, the difference in development from region to region should be kept within certain limits, and the government needs to take special measures to spur the development of underdeveloped regions and prevent the gap between the rich and the poor caused by the cumulative and causal cycle from widening indefinitely.

From the perspective of the relationship between the central government and local governments, institutional economics believes that due to the universal character of moral hazard, although the local governments and central government are the same in nature, there are differences in interests between the two. The central government represents the country as a whole and shoulders the responsibility of safeguarding sustainable economic and social development by giving priority to the overall interests of the country. The local governments, however, would pursue local interests and self-interests at the expense of the overall interests when the former are large enough for them to break through the constraints of government norms, thus rendering the entire social and economic life be caught in the serious "tragedy of the commons" under new conditions. Therefore, intervention has a point.

As for the relationship between local governments, their competition is of an exclusive nature, which means each one of them tries to maintain its own "turf" by instinct. From the perspective of expansion, once the natural factor of physical locality is fixed, the way for local governments to expand outward is by growing its economic power outside its territory and thus gaining more share of the cake. Since every local government intends to do so, the attempt to increase the share of the cake is achieved by way of population game. According to new institutional economics, the two parties in conflict often resort to a third party to solve problems, such as the transaction cost is too high, or worse, the failure to transact affects the resolution of conflict. The coercive power or influence of the third party is used to force one of the parties to settle the issue through equal consultation. The introduction of third parties thus rendered conflicts that would otherwise be impossible to resolve through negotiation resolved, turning non-cooperative games into cooperative games.

As to the relationship between developed regions and underdeveloped regions, economic disparity is only the surface of development inequality, and the roots hidden deeper is the inequality of political power and the lack of development capacity. The ethical choice that a society should make must consider the opportunities afforded the most unfortunate population in society, and only a society of equal opportunities is a fair society. Therefore, the provision of equitable public services is the key to redress the backwardness fundamentally.

2.3 Theoretical Framework of China's Urban Spatial Development

2.3.1 Fundamental Perspectives

Considering the development stage of urbanization, the resource and environment conditions available to urban development, and the basic planning system currently in use in China, the following basic principles need to be acknowledged in the theory of urban spatial planning at the national level:

(1) View on Global Development: The national spatial strategy must be tied to global economic and social development. To enable cities in different regions to integrate into the world economic trend, make effective use of international markets and international capital, take full advantage of international resources (energy, technology, culture, management) and promote national economic development; to build a globally oriented spatial structure (global network, node cities and regions), integrate into the world city system and support the development of Mega-City Region, gateway cities and regions with high concentrations of economic resources.
(2) View on Resource Constraints: To focus on the constraints of resources and environment on spatial development and take the path of sustainable development. Guided by the science of human settlements, to integrate the five systems—nature, human, society, residence, and support [6], to determine the master plan and specific policies for human settlement development from the perspective of the scientific and effective use of spatial resources, and to develop a spatial system characterized by ecological security, economic efficiency, and urban–rural coordination.
(3) View on Spatial Dynamics: To embrace the trend of global development, focus on the development priorities at the current stage and arrive at an open and flexible urban spatial structure. To create an analysis system for the correlation between "economic space of flows" and "human settlement space", enhance the resilience of spatial structure, monitor the trend of regional spatial development, and adjust the priorities of spatial development in a timely manner. To improve the services of regional transportation infrastructure and provide favorable conditions for spatial growth.
(4) View on Equity in Services: To build a standard system of public service facilities that integrates urban and rural areas under the principle of equity in services. To lay down the requirements for establishing public service facilities of education, health, culture and so on, and propose supportive policies for spatial development, especially financial and tax policies. Equity between regions should be more manifested in the equalization of public services.

(5) View on Government Intervention: The central government intervenes in the socioeconomic development of the entire country at the macro level through national spatial planning, remedies the shortcomings of market through macroeconomic regulation, reflects the political belief of fairness, deals with the intricate relationship between the central and local authorities and among local authorities, determines policy interventions for spatial development at different levels, protects precious resources that have a bearing on national interests and the safeguarding of the quality of overall human settlements, and guarantees the coordinated development of economy, society and environment.

2.3.2 Planning Methodology

(1) **Identify Cities and Regions with International Competitiveness in Response to Economic Globalization**

First, analyze international capital flows and gain insight into the development trend of priority cities and regions. A thorough analysis of the interactions between capital flows, industrial linkage, regional collaboration, and urban space in China in the context of globalization is the basis for the analysis of the country's urban spatial structure. Given the trend of international investment continuing to pour into the east, a spatial policy for the coastal Mega-City Region should be established as soon as possible to enhance the country's ability to participate in global competition. Attention should be paid to the demands of large industries on space and their impact on environment, and the pressure from dense structures and environment degradation should be alleviated. At the management level, the planning and management mechanism for the Mega-City Regions centering around Shanghai, Beijing, and Guangzhou are to be established to guarantee the overall integrity and environmental quality of those regions.

Second, analyze the impact of regional economic cooperation organizations and enhance the development of node cities and cross-border transit routes. The establishment of the urban spatial structure in China, a country located in the Asia–Pacific region, should be largely aimed at the common development of the Asia–Pacific region. The focus should be on the development of cross-border transit routes and the building of a transport corridor gearing towards the Asia–Pacific region. New growth poles should be cultivated based on the core cities in the border areas and cross-border economic cooperation zones should be established in the northeast with Harbin as the center, in the southwest with Nanning and Kunming as the center, and in the northwest with Urumqi as the center, respectively. Another focus is the development along the oil import corridors covering Russia, Kazakhstan, the Middle East, and southeast Asia. A holistic approach should be taken to the common development and management of water resources in the Mekong River basin. The development of logistics ports based on small- and medium-sized border cities should also be strengthened.

Third, fully recognize the role of industry cluster and guide the operation of small- and medium-sized cities and small towns by taking advantage of the situation. Given the pace of economic globalization and the huge agricultural population in China, boosting the development of industrial clusters featuring specialty industry is not only an important means to embrace the global shift of industries, but also a critical measure to enhance the relevance of township and village enterprises. To cultivate the core competitiveness of small- and medium-sized cities and re-create the development momentum for small towns is an important part in achieving sound urbanization and promoting coordinated development of urban and rural areas.

(2) **Making Decisions about the Protection and Management of Precious Resources at Different Levels Based on Sustainable Urban Development**

First, analyze the layers of spatial resources based on ecological security. The spatial resources are divided into ecological security layer, infrastructure layer and human settlement layer, and the analysis of the three layers is used as the basis for constructing a planning method for the sound development of urban space. The ecological security layer refers to the spatial resources based on natural and ecological elements, which is the foundation of human settlements. The infrastructure layer refers to the infrastructure network with transportation as the core, which constitutes the backbone of space. The human settlement layer refers to the human settlements of various types and at different levels on which human life revolve, such as cities, towns, villages, and the living quarters around industrial mines. The analysis of the spatial resources at the national, regional and city levels is the prerequisite for sound planning.

Second, establish a three-level spatial resource planning and management system. It is imperative to establish a three-level planning and management system for national, regional, and urban planning with resource conservation (natural and cultural) as the main content based on the three-level analysis of spatial resources. Inventories of resource protection at different levels should be formulated, with a focus on proposing measures and means of protection at the national and regional levels. In terms of resource protection, the national-level spatial planning should focus on natural resources, including water resource areas, forest resource areas, wetland areas, and the areas prioritized for ecological restoration, as well as historical and cultural resources, both at regional level and in towns and cities.

(3) **Scientific Prediction of Population Growth and Movement and Accurate Identification of Priority Areas for Urbanization**

First, establish a gray prediction model of industrial development, population movement and spatial expansion. The evolution of urban spatial structure is a process of continuous adjustment amid population movement and industrial development. Strengthening the analysis of the relationship between population, industry and spatial change is the key to accurately grasping the trend of urbanization. It is necessary to predict the trend of cross-regional movement of labor force, assess industrial demand on land in cities, and determine the development scale and direction of regional cities considering the changing job market brought about by industrial development.

Second, identify priority areas for urbanization based on the general trend of population movement. With the development of regional economy and the widening of urban–rural disparities, the gradual movement of population to economically developed regions and urban areas is an inevitable result. And with the growing economic influence of central cities, the development of urban clusters around central cities is gaining a positive momentum. The status of population concentration showed that the Pearl River Delta, the Yangtze River Delta, and Beijing-Tianjin-Hebei region are the secondary densely populated areas. Given the trend of future development, the regions with faster economic growth, including south-central Liaoning, Shandong Peninsula, southeast Fujian, Wuhan, central Shaanxi (Xi'an), Central Plains (Zhengzhou), Changsha-Zhuzhou-Xiangtan, Chengdu-Chongqing, Changchun-Jilin, and Beibuwan, will see high concentrations of the floating population, and thus become prioritized for urbanization.

(4) **Establish A New Regional Coordination Mechanism for Coordinated Regional Development**

First, establish urbanization policy zoning around regional central cities. Regional policy is one of the core elements of national spatial planning. The policy zoning of urbanization should consider both the division requirements of comprehensive economic zones and the characteristics of regional urbanization. Since the Reform and Opening Up, China has introduced a series of comprehensive programs in response to the changing needs of different times, from the trailblazing development of the eastern region, the large-scale development of the western region, the revitalization of the old industrial bases in the northeast to the rise of the central region. Therefore, the following principles should be observed: consider the consistency of regional geographical features and maintain the continuity of regional history and culture; consider the closeness of social and economic ties within the region and the scope of influence of economic hub cities; maintain the integrity of provincial and regional administrative units as much as possible; consider the complete network structure of regional infrastructure and the role of regional hub cities. Use one to two mega-cities as the core to organize regional economic activities, and as the gateway for the region to participate in global competition. This way, both regional economic activities are organized, and the requirements of urbanization policy implemented. Seven new urbanization policy zones are proposed, and different policy requirements are put forward for each of them. Taking North China for instance: improve the quality of urbanization in Beijing-Tianjin-Hebei region, develop the Shandong Peninsula city cluster proactively, and foster the development of city and town clusters around Beijing and Tianjin in Hebei Province; improve the level of industrial modernization, develop high-tech industries and tertiary industries actively, and cultivate Tianjin Binhai New Area as a new economic growth pole; improve the collaboration between Beijing-Tianjin and Hebei, Shanxi and Inner Mongolia on energy utilization and environmental protection, and protect the environment of Inner Mongolia and northern Hebei in strict compliance with laws and regulations;

strengthen the construction of cross-provincial transportation corridors, and coordinate the operation and management of the ports in Bohai Bay, including Tianjin Port, Tangshan Port (Hebei), Qinhuangdao Port (Hebei), Huanghua Port (Hebei), Qingdao Port (Shandong), and Yantai Port (Shandong).

Second, establish a unified and efficient public service system and promote urban–rural integration. The coordinated development of urban and rural areas, the biggest challenge facing China's urbanization at present, is both a necessary part of building the new socialist countryside and a prerequisite for improving the quality of urbanization. Given the realities in China, the development of urban and rural infrastructure should focus on improving transportation and municipal utilities, and giving priority to the construction of road networks linking major towns and villages; popularizing and promoting nine-year compulsory education in rural areas, expanding the coverage of high-quality educational resources, and making extra effort to promote vocational education and adult education in areas with major labor inflows and outflows; establishing the three major systems of medical services, preventive health care, and health supervision, establishing and improving the rural cooperative medical care system, and creating a health security facility system covering both urban and rural areas; promoting the development of non-profit cultural programs, facilitating the prosperity of for-profit cultural industry, and building a relatively complete system of public cultural facilities of different levels to meet the needs of different groups in the society; gradually establishing a social welfare management system and operating mechanism compatible with the socialist market economy, and accelerating the development of social welfare and relief facilities such as pension service facilities, children's welfare services, social relief facilities, service facilities for people with disabilities and funeral service facilities.

(5) **Build A Dynamic and Diverse Urban Spatial Structure Based on the Uncertainty of Development**

First, develop a globally oriented and dynamic urban spatial structure. It is imperative to link the urban system of China up with that of the world as soon as possible, especially to integrate with the urban space of the Asia–Pacific region and create an open spatial structure. Several cities with international influence are to be chosen for global competition and developed into international cities comparable with New York, London, Paris, and Tokyo. The Yangtze River Delta with Shanghai as the core and the Pearl River Delta with Hong Kong as the core are the most eligible to be chosen and built into world-class cities. Carry out dynamic multi-scenario simulation according to the different stages of industrialization and urbanization in China and their development trends. With the evolution of industrialization, the national spatial structure plan should be adjusted to suit the changing situation at different points in time.

Second, build the main body of urban spatial structure centering around urban cluster. Urban cluster is a manifestation of regional urbanization and an advanced form of urbanization. With the progress of urbanization, urban cluster, a unique urban spatial form, is bound to play an increasingly important role in urbanization in future given its huge economic impact and remarkable space efficiency. Building the national urban spatial structure with urban clusters as the mainstay is in line with the trend of urbanization and conducive to the intensive use of space, which would play an effective role in promoting coordinated regional development and improving the overall competitiveness of the country. 16 urban clusters are marked for priority development considering the current conditions and potential for development.

Third, establish support system and service system for urban development with transportation as the core. The development level of urban space depends largely on the supply of infrastructure. In terms of national spatial development, whether less developed regions can prevent themselves from being marginalized hinges on the accessibility of their locations and the degree of connection between its regional economy and that of the developed regions and even the world economy brought about by this accessibility. Building a hierarchical regional transportation network covering all major spaces in the human settlements is a crucial part of a sound national spatial structure. Transport hubs shall be constructed in north China, east China, south China, central China, northeast China, northwest China and southwest China with Beijing, Shanghai, Guangzhou, Wuhan, Shenyang, Xi'an, and Chengdu as the centers, respectively.

(6) **Promote the Transformation of Urban Economic Growth Guided by New Industrialization**

First, develop the concept of new industrialization in response to the pressure and challenge from development. Industrialization is a development stage a country cannot skip in the process of modernization, and a specific historical stage and process of transition from an agricultural-based economy to an industrial-based economy. Under the influence of the new technological revolution, the concept of economic development centering around traditional industries is facing unprecedented challenges. The characteristics of new industrialization include: to drive industrialization by informatization and vice versa; to improve the quality and efficiency of economic growth by advancing scientific and technological progress and improving the quality of the labor force; to take the path of sustainable development by attaching importance to the balance between economic development and ecological conservation; to give full play to the advantages of human resources by properly dealing with the relationship between improving productivity and expanding employment.

Second, accelerate the adjustment of industrial structure and update the urban and regional industrial structures. To achieve the development goals of new industrialization, we should first deepen the industrialization process nationwide and promote coordinated development of the eastern, the central, and the western regions. Second, we should rationalize and modernize industrial and economic structures and develop a more sophisticated industrial structure. Third, we should accelerate the industrialization of agriculture and properly deal with the relationship between industry and agriculture. Fourth, we should transform the mode of industrial and economic growth, pursuing intensive economic growth and improving economic efficiency through technological progress. Fifth, in terms of the implementation mechanism, we should give a bigger role to market mechanism, to effectively transform the functions of government. Overall, China is at the beginning of the middle stage of industrialization with great disparities in the industrial structures among different regions. In the central and western regions, agriculture, and light industry, which had played a major role in the early stage of industrialization, need to be further developed. In the eastern and central region, heavy chemical industry, which had played a leading role in the middle stage of industrialization, will be enhanced. In the eastern coastal areas, emerging industries, which have been playing a significant role in the late stage of industrialization, will thrive.

(7) **Establish the Planning System Based on Efficient Spatial Management and Integration of the Three Plans into One**

First, establish the regulatory system for coordinated socioeconomic development with spatial planning as the mainstay. In a market economy, economic activities are mainly regulated by the market. Since China's establishing the socialist market economy in 1992, and especially after her accession to the World Trade Organization in 2003, China has become a market economy in development with the perfection of its market system. Industrial development is basically regulated by the market, and even the formulation of major national industrial policies mostly considers the needs of the international and domestic markets. In countries with relatively complete market economic systems, such as Japan, South Korea and EU members, the means they used to regulate economic development are spatial policies and spatial planning. Therefore, we should establish a spatial planning and regulation mechanism that conforms to our national conditions, shifting from the traditional "national economic and social development plan" with notable vestiges of the planned economy to the "national spatial planning" centering around the sound allocation of spatial resources.

Second, establish a unified and efficient spatial plan formulation and management system integrating the three plans. At present, the spatial development plans at the national level mainly include the *Five-Year Plan* for national economic and social development" formulated by the National Development and Reform Commission, the "National Urban System Plan" prepared by the Ministry of Housing and Urban–Rural Development, the "National Overall Land Use Planning" formulated by the then Ministry of Land and Resources as well as various special plans prepared by numerous ministries and commissions including transportation and environmental protection. As different plans are the charge of different departments, and the preparation time and duration of them are different, there are many conflicts between them. It is important to give a thorough overhaul to the current planning agencies under different ministries and commissions (especially the National Development and Reform Commission, the Ministry of Housing and Urban–Rural Development and the then Ministry of Land and Resources) and establish a national spatial planning management authority led by a comprehensive department and a planning system that "integrates the three plans" to ensure the rational use of spatial resources in China.

Third, establish plan implementation measures and instruments based on economic policies including finance and taxation. Basically, the experience of other countries in spatial plan implementation indicates that fiscal and taxation instruments are the most important measures. The coordinated spatial development of

China should focus on the following: first, establish a "national regional development fund" to support the development of less developed regions and the revitalization of declining regions; second, use national fiscal policy such as mobile payment to promote the protection and sound use of resources in sensitive regions; third, use monetary policies to give full play to the role of policy banks and regional banks to promote local economic development; fourth, develop regional labor and employment policies to promote the sound flow and optimal allocation of labor resources between regions while preventing the loss of knowledge-based labor force in less developed regions and promoting coordinated development of regions.

(8) **Establish the Think Tank for Urban Spatial Planning and Facilitate Sound Decision-Making on Urban Planning**

First, establish the Special Committee on Spatial Planning of the National People's Congress to Regulate Planning Legislation. A "special committee on spatial planning" should be established in the National People's Congress to sort out the plans prepared by various departments under the State Council and integrate them from the perspective of legislation. In the United Kingdom, Japan, Singapore, and the United States, the planning committees established at different levels are composed of planning professionals, who can put forward policy suggestions on planning legislation from the perspective of a third party, thus promoting sound decision-making.

Second, improve the national and provincial planning inspection systems to facilitate sound decision-making on regional space. Considering China's political system and national conditions, the establishment of a top-down planning inspection system with professionals and technicians as the backbone is an important way to guarantee sound spatial planning decisions and implementation. The specific duties of the national planning inspectors mainly include supervising the implementation of the national urban spatial planning, propose major construction projects across regions, and provide coordination on cross-provincial matters, etc.

Third, establish the position of chief planner at provincial and municipal levels to facilitate and supervise the implementation of national spatial planning. To establish the position of chief planner in provinces (autonomous regions) based on city chief planners is an important means to guarantee sound decision-making of provincial governments on planning. The professional services can, to a certain extent, correct the mistakes in the decision-making of the "tenure system". The relatively stable system of provincial (autonomous regions) chief planner is effective in implementing the medium- to long-term national urban spatial planning (Fig. 2.5).

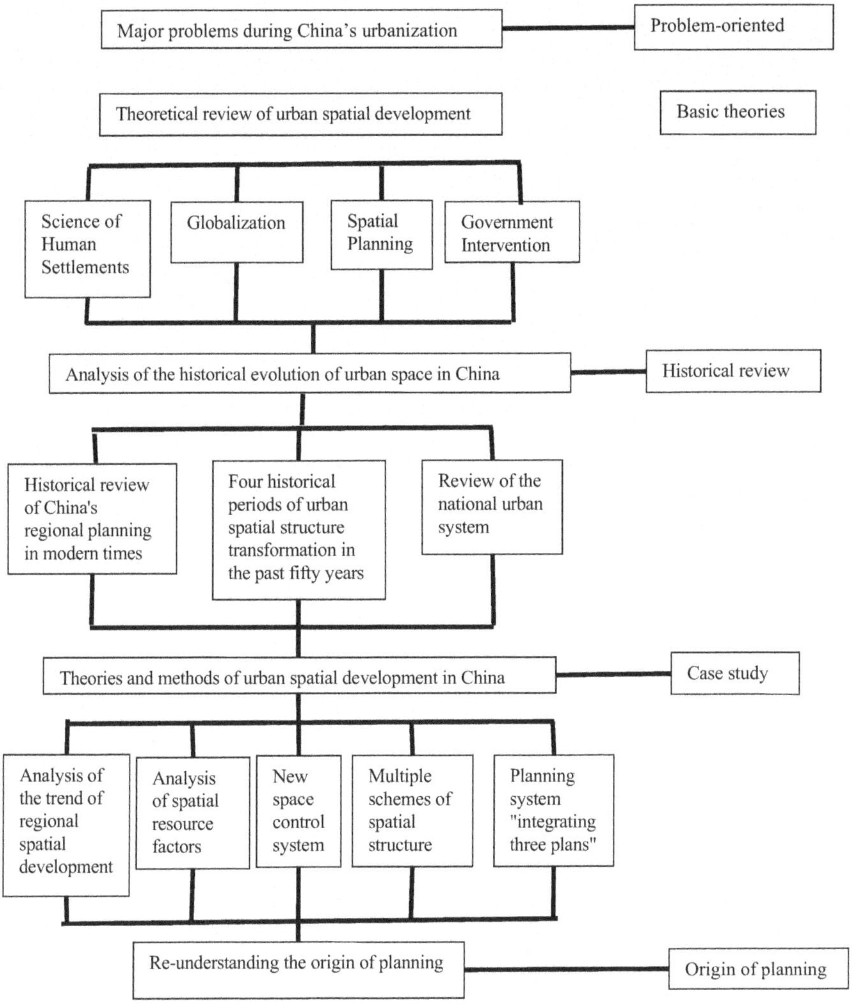

Fig. 2.5 Diagram of study structure

References

1. Yang WM (2003) Theoretical exploration of planning system reform. China Market Press, Beijing, p 6
2. Wu CF et al (1995) Introduction to land science, vol 8. China Building Materials Press, Beijing, p 23
3. Li Y et al (1999) Land resources and economic layout. Geology Press, Beijing
4. Cui GH et al (1999) Regional analysis and planning, vol 1. Higher Education Press, Beijing, pp 345–346, 353
5. Wu L (1982) Rehabilitating the Old City of Beijing: A Project in the Ju'er Hutong Neighbourhood. University of British Columbia Press. https://muse.jhu.edu/book/41888

6. Wu LY (2001) Introduction to sciences of human settlements, vol 70. China Architecture and Building Press 101(106–112):183

Chapter 3
Theoretical Review of Urban Spatial Development

3.1 Analysis of Comprehensive and Coordinated Urban Development According to the Science of Human Settlements

3.1.1 Theoretical Context

The science of human settlements advocated by Mr. Wu Liangyong is important for analyzing and understanding the law of urban spatial development and making sound urban spatial planning at the macro level. According to the theory, human settlements refer to the human-centered living environment that covers both man and other living beings and people's living systems. Any human settlement is founded on the following five aspects: ecology, economy, technology, society and human and culture. Drawing on the idea of systems science, the theory proposes a new methodology for the study of urban spatial development at different levels, believing that human settlement is a complex giant system composed of the five levels of the globe, regions, cities, communities (villages and towns), and architectures, which demands a coherent and comprehensive research method, and that the study of cities and regions should be regarded as a science about the whole and wholeness. The discourse of spatial planning at the regional (including national) level focused on the experience of the Netherlands, stating that there is a certain correlation between the resource conditions and population density of the country on the one hand, and the impact of national policies on the region on the other.

As far as China's urban development is concerned, this theory believes that it is necessary to start with the national conditions, grasp the imbalance of its urbanization progress, and explore different paths of urban development. For the urban system, certain patterns can be identified in the evolution of urban forms. For instance, mega cities and large cities tend to shift from centralized and monocentric structures to dispersed and regional structures. In economically developed and densely populated

K. Wang, *National Spatial Planning in China*, Urban Sustainability,
https://doi.org/10.1007/978-981-97-7729-7_3

areas, rural towns tend to become relatively concentrated with the spread of industry. For the entire urban system, "large, medium, and small cities need to develop in a coordinated manner to form a sound urban system, and a comprehensive network between urban and rural areas and between regions need to be established gradually to promote coordinated economic, social and cultural development of urban and rural areas" [1].

3.1.2 Enlightenment of Planning Practice

Since the 1980s, Mr. Wu Liangyong has done systematic research on urban spatial development at the regional level, starting from "the development and planned construction of small towns in the Taihu Lake area". In the study of the overall urban planning of Shanghai and Suzhou in the 1980s, he proposed to construct spatial structure of cities from the perspective of regions, such as Shanghai's strategy of Pudong development and Suzhou's "Big Cross" structure. He also voiced his views on such important issues as the development of the Mega-City Region and the integration of developed regions, holding that the Mega-City Region should seek to achieve integrity in the following four aspects: economic development, regional space, urban and rural development, and stage of development. In the research of "Protection and Development of the Architectural Environment in the Process of Urbanization in Developed Regions", he conducted in-depth research on the urban spatial development of Shanghai-Nanjing-Hangzhou area, proposing to build not only compact cities, but also areas of sustainable development and regional infrastructure networks with systematic and open characteristics. It is imperative to deepen the understanding of the systematization of regional development; attach importance to the integrated research while taking account of the limited goals; and comprehensively analyze the relationship between regional economic, social, environmental, and cultural development and spatial region, as well as the relationship between the overall coordination of regional development and regional infrastructure corridors, emerging localities of regional spatial development and their spatial forms, i.e. to promote regional planning from the perspective of urban planning (Fig. 3.1).

In the "Research on the Sustainable Development Planning of Human Settlements in Northwest Yunnan[1]" later, he focused on the construction of human settlements under complex ecological and living conditions. In the "Research on Human Settlements in the Three Gorges Reservoir Area[2]", he put forward a series of important views on natural environment, protection of human and culture resources, urban development, and regional industrial structure adjustment in large-scale project

[1] See "Research on the Sustainable Development Planning of Human Settlements in Northwest Yunnan", Wu Liangyong. Kunming: Yunnan University Press, 2000.

[2] See "Research on Human Settlements in the Three Gorges Reservoir Area", Zhao Wanmin. China Architecture & Building Press, 2000.

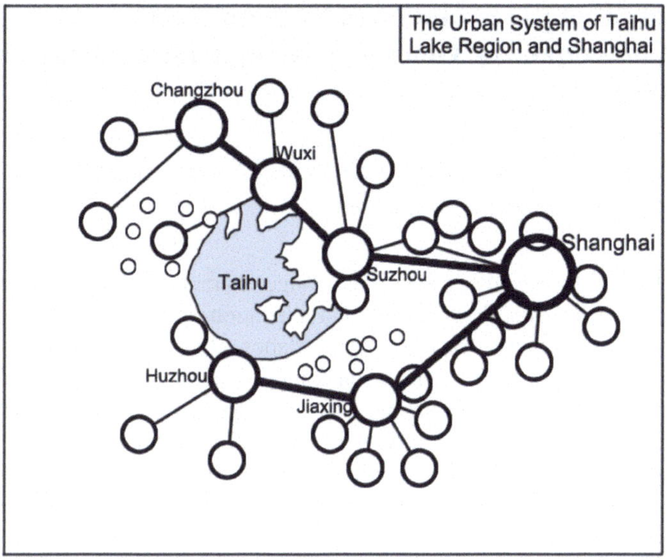

Fig. 3.1 Diagram of the "Comet-Shaped" urban system in Shanghai and Taihu Lake. *Source* "Construction of Small Towns in Taihu Lake Area", Wu Liangyong

construction. In the "Research on Urban and Rural Spatial Development Planning in Beijing-Tianjin-Hebei Region", he further proposed the important idea of constructing urban spatial structure from the perspective of regions considering the development of Mega-City Region under globalization. Going beyond the administrative confines of Beijing, the study proposed the spatial structure of Greater Beijing based on the overall coordinated development of Beijing, Tianjin, and Hebei, which has long-term strategic significance for the adjustment of urban spatial structure of both Beijing and Tianjin and clarifies the thinking on urban development of Hebei. It can be seen from the important planning practices above that the science of human settlements believes in the development of urban space it is imperative to consider such external factors affecting the human settlements as the ecological environment, economic foundation, social and cultural condition. In the complex surroundings, it is important to identify and solve the main problems considering the characteristics of the region, and the framework of the entire research and planning is a complete network system that includes problem analysis, goal setting, sorting out the action plan, and specific measures for implementation[3].

[3] Starting from the research on "Protection and Development of the Architectural Environment in the Process of Urbanization in Developed Areas" and in the research of Northwest Yunnan and the Three Gorges Reservoir Area, Mr. Wu advocated a coherent and holistic research framework, proposing the schematic matrix diagram with different themes on many occasions.

3.2 Analysis of the Evolution of Regional Space in Contemporary Times According to the Globalization Theory

3.2.1 Basic Understanding of Regional Space According to the Globalization Theory

Since the 1970s, the world has seen the change characterized by economic globalization, accelerated free flow and optimal allocation of production factors across the globe, deepened international division of labor, globalized production process, increasingly closer economic ties between countries and regions, and the contemporaneous strengthening of the following two trends at the urban and regional levels: re-decentralization of production and re-centralization of management [2]. At the institutional level, market rules and political laws are also becoming increasingly global [3]. Globalization has become an important concept widely utilized by the academic community in the analysis of global economy and social development, on which the planning community and geographical community have done a lot of research, especially the process of geo-spatial change and its significance in the context of globalization.

Theoretically, economic globalization means the worldwide spread of industrial production, commercial activities, capital flows, communication technology, information dissemination, currency, etc., disregarding the existence of borders, that is, the so-called "advent of a borderless world". It also means that almost all economic activities are connecting places through networks, the rapid mixing and merging of cultures in both form and content, and their becoming more dependent on the world market [3]. It is safe to say that local economic activities are increasingly involved in the restructuring of production and consumption activities on a global scale [4]. Although some scholars disputed the "death of spatial distance" after the events of "9/11" in the United States in 2001 [5], the consensus of governments is that economic globalization remains the mainstream of future development[4].

The theory of globalization holds that the world economy has undergone a new round of reorganization due to globalization, which is characterized by the following three aspects: first, the strengthened centrality of the global financial structure. Financial capital has become an independent force in modern society, as evidenced by the fact that currency has exceeded all spatial and location limits, and its role in the allocation of global resources has become ever stronger, creating a new monetary geographical space. Second, transnational corporations have attained an enhanced status in international economic exchanges and national power. Amin [3] argues that we have entered the new era in which transnational corporations have the power to bargain with national governments at the global level; it is through those corporations that the division of labor in global industries is largely realized. Third, administration

[4] In his report to the Third Plenary Session of the 16th CPC Central Committee, Hu Jintao held that economic globalization will continue to deepen in the next 20 years.

at the national level was weakened while that at the regional level has been strengthened. Since capital has stronger motive to flow across borders and thus become more regional in character, the correlation between vertical centralized management at the national level and market sharing is weakened, and cooperation between regions and global markets is strengthened.

Globally, the flowing of capitals has brought about the worldwide relocation of productive space, i.e., the emergence of the so-called "space of flows", which has been manifested mainly by foreign direct investment (FDI). For nearly 20 years, the cross-border flow of long-term capital has been at the heart of global business development [6]. FDI showed a dramatic upward trend from the early 1980s to the late 1990s. From 1981 to 1985, the average annual flow of foreign capital was 48 billion US dollars. The figure rose to 186 billion between 1991 and 1993 and further to 735.1 billion in 2001 [7]. The proportion of foreign capital flowing into developing countries including China kept increasing, from an average of 74.3 billion CNY per year from 1990 to 1995 to 204.8 billion CNY in 2001. Alden [8] believes that throughout the world, foreign investment is playing its role at all levels of space and featuring prominently in economic development policies of various countries,[5] it serves as a great driver for local economic development.

Economic globalization has been accompanied by trade liberalization. Beginning in the late 1980s, trade liberalization across the world has been accelerating. According to statistics, the growth rate of international trade in the 1990s was almost twice that of the world production, and most of the growth came from exports from developing countries [6]. These exports have something to do with the process by which transnational corporations place production in different countries to minimize the cost of different stages of the production process, i.e., the so-called "slicing up the value chain". It is estimated that almost one-third of the world trade in the mid-to-late 1990s took place in such global production networks (World Bank, 2000). It can be said that the liberalization of trade promotes the globalization of production, and vice versa, and the interaction between the two has brought about the expansion of many industrial spaces in developing countries and the wave of urbanization and the tremendous development of urban space as a result.

3.2.2 Impact of Transnational Corporations on Regional Space

As mentioned earlier, the main feature of economic globalization is the global flow of foreign capital. The flow of foreign capital in the past 50 years has been globalization at the enterprise level represented by transnational corporations. Being the major players in international trade, transnational corporations are both the exporters and

[5] Since 1979, the Chinese government has been drawing foreign capital by establishing special economic zones and economic and technological development zones, which have had a significant impact on the country's economic and social development.

importers of final goods and services; they are a major source of innovation in the world economy [6].

According to the definition of the Organization for Economic Cooperation and Development (OECD), transnational corporations refer to the enterprises with production facilities in two or more countries. And the *Investment Report 1996* of UNCTAD defined a transnational corporation as an enterprise that owns a certain amount of equity capital and thus controls the assets owned by foreign economic entities [9].

As an economic group, transnational corporations are the most dynamic organizations in the world. Statistics showed that in 2001, there were 65,000 transnational corporations in the world, with 850,000 foreign affiliates, staff of 54 million, and sales of 19 trillion US dollars. Many a transnational corporation is comparable to a country in terms of wealth, and 29 of the world's top 100 largest economic entities are transnational corporations [9]. According to statistics, in 2001, foreign affiliates of transnational corporations generated 11% of the world's GDP and 1/3 of global exports (UNCTAD, 2002), which means the affiliates located in developing countries are already the mainstay of the development of transnational corporations.

Research shows that since the 1990s, large companies in a wide range of economic sectors of the world have become increasingly concentrated, such sectors including aviation, automotive, pharmaceuticals, petroleum processing, banking, insurance, advertising, and mass media. The data show that the top companies are getting more and more shares of the global market with a handful of companies taking as much as half of the market. It can be said that commodity supply and service agglomeration is moving towards system integration at the global level [6],the production space is becoming increasingly fragmented worldwide, and at the same time, ever more concentrated in certain areas according to the specific category of industry the enterprises invested in [4] (Table 3.1).

The rapid development of transnational corporations has a great impact on the form of global space, and the progress of information technology lent new technical support to this change, making it possible for people to effectively control the existing spatial organizations across regions, especially the production organization of the

Table 3.1 Trends in global FDI inflows over the past 10 years unit: billion US$

	1990–1995 Average	1996	1997	1998	1999	2000	2001
Global	2253	3861	4781	6945	10883	14,919	7,351
Developed countries/ regions	1450	2199	2679	4842	8378	12,275	5031
Developing countries/ regions	743	1527	1910	1876	2251	2379	2048
China	194	402	442	438	403	408	468

Source 2002/2003 Report of Transnational Corporations Investment in China, 2003.
UNCTAD World Investment Report 2002

secondary industry. Therefore, it is only natural that investments are pouring in developing countries of relatively low labor costs and land costs. Since transnational corporations are at an advantage economically, they are the ones taking initiative in their investment in underdeveloped countries, which, if reflected in space, means they tend to choose development zones with better technical infrastructure and cheap land prices in large and medium-sized cities, and its characteristics of being large-scale and "top-down" are obvious.

Paul Krugeman [10] believed that the effect of increasing returns is to make each product profitable only when produced in one place, and the result is that different products are produced in different places. From the perspective of economic globalization, it means production is increasingly taking place throughout the world, yet at the same time it is ever more concentrated at certain regional places. This also largely explains why industries (manufacturing in particular) keep gathering in the developed regions of some countries, and accounts for the emerging regional imbalances brought about by the imbalance of industrial spaces within a country.

3.2.3 Impact of Industrial Cluster on Regional Space

Coexisting with the global industrial arrangement of transnational corporations is the following reality: enterprise clusters or industrial clusters mainly composed of small and medium-sized enterprises have gradually become a crucial component of the regional economic development of various countries. Being a common phenomenon, the reality with the advocacy of relevant organizations such as the United Nations, industrial cluster has gradually become an important part of the public policies of various countries [4].

The so-called "industrial cluster" refers to a group of interconnected companies and related organizations that are geographically close to each other and are in the same or related to a particular industrial sector, and they are linked because they have something in common and are complementary to each other [11]. An industrial cluster includes a host of interconnected industries and other entities that are vital to competition, and it is often extended downward to sales channels, clients, manufacturers of supporting products, and industrial companies related to technologies or inputs. The concept first appeared in Porter's *The Competitive Advantage of Nations* from 1990.

Industrial cluster is a common sight around the world. According to Wang Jici's research [4], major representatives are the microelectronics clusters of the Silicon Valley and Route 128, the advertising industrial cluster of Madison Avenue in New York, the medical equipment industrial cluster in Minneapolis, the entertainment industrial cluster in California, and the footwear industrial cluster in Massachusetts. In Italy, more than 70% of manufacturing industry, more than 30% of employment and more than 40% of exports are realized in specialized industrial areas. In Germany and France, similar examples abound, such as the cutting tool industrial cluster, the machinery industrial cluster and the network industrial cluster, and the perfume glass bottle industrial cluster. In Peru, Brazil, Mexico, and other Latin American countries, there are hundreds of regional governments and about 15,000 cities, and cluster programs are implemented almost everywhere. In Asia, there are specialized clusters with different degrees of development in Japan, India, South Korea, Pakistan, and Indonesia. The economic output of clusters has accounted for a high proportion of the national economy. Taking California in Silicon Valley for instance, its economic aggregate ranked it the 11th in the list of total economic volume by countries. The prevalent clusters have formed "economic mosaic" with colorful and distinct blocks (Wang Jici, 2003).

Enterprise clusters possess apparent external scale effects and strong industrial correlation. In terms of regional space, it enjoys a considerable degree of self-organization, forms relations purely by the industry chain, and concentrates in cities and towns, thus forming a "cluster" of industries and spatial agglomeration. Its characteristic of being "bottom-up" is obvious. This is true of the cluster along Route 128 in the United States, as well as the garment industrial cluster in Ningbo, Zhejiang.

3.2.4 New Urban Hierarchy around the World

So far, we have analyzed the reasons of regional spatial change due to the characteristics of economic activities. In fact, the hierarchical relationship of economic activities itself also has an impact on the structure and content of the urban system. Hall [12] has done an analysis of the functions of global cities and concluded that London, Paris, New York, and Tokyo are the typical representatives of global cities, whose main functions are finance and commerce (including banking, insurance, law, design, etc.), command and control (national governments, international organizations, headquarters of transnational corporations), culture and media (performances, exhibitions, museums, printing), and tourism (hotels, restaurants, entertainment, transportation, etc.). Since these high-end services and activities are highly interconnected, it is inevitable that they are gathering in certain cities and regions.

Sassen [13] argues that over the past 15 years, the focus of economic activities has largely shifted from manufacturing bases such as Detroit and Manchester to financial and high-end specialized service centers, including both the well-known "global cities" and the so-called "sub-global cities.“ Hall [2] did a holistic analysis of the functional attributes of cities and the interrelationships between cities and revised the traditional city hierarchy system of Christaller, proposing a new urban hierarchical structure of global cities, sub-global cities, regional cities, and provincial cities.

Mega-city regions are not monocentric spatially and functionally, but multi-centric and continuous in terms of space. This type of region is not only the natural result of economic activities, but also serves as a "new cradle of space organization of human settlements" [14]. Typical examples include the southeast of England, the New York area, the Tokyo-Osaka area, and the Yangtze River Delta and Pearl River Delta in China.

3.2.5 Spatial Structure of the Asia–Pacific Region in the Processes of Deconstructing and Reconstructing

When analyzing the changes in space organization worldwide, the changing process of the Asia–Pacific region in the past few decades deserves special attention. First, the Asia–Pacific region has experienced rapid industrialization over the past 50 years. From the global perspective, this process can be seen as an important re-agglomeration of global industries. Some studies believe that despite the prolonged period of the world economic recession since the 1980s, the Asia–Pacific region has basically maintained its development thanks to the reforms of China and other former planned economies. The well known "Swan Chart" shows the economic development of the Asia–Pacific region and its integration into the world economy in the past five decades. This process of change was taking place at various levels include the globe, regions, and countries (Fig. 3.2).

Some studies suggest that urbanization in the region can be categorized as investment-induced "exo-urbanisation", Desakota [15], and urbanization driven by large-scale programs [2]. McGee's desakota has made considerable influence, and building on this research, McGee went on developing the concept of "extended metropolitan regions", which is supported by such cases as Bangkok and Jakarta. According to McGee [16], urbanization in the Asia–Pacific region is driven by three main forces: globalization, transactional revolution, and structural change. The characteristics are not only new urban forms, but also a new type of transnational economic development manifested in space.

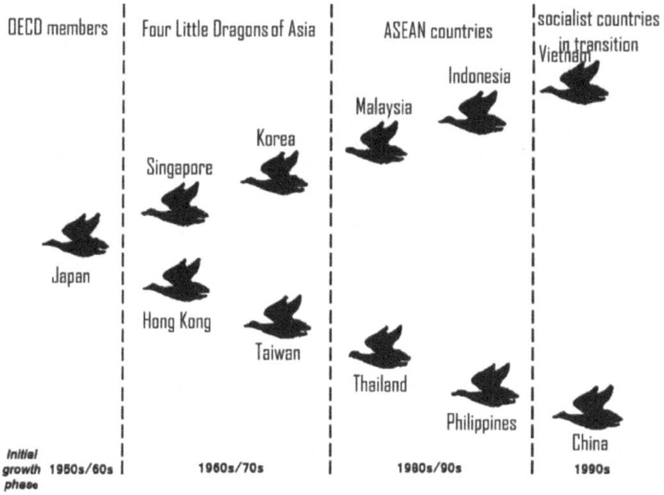

Fig. 3.2 Diagram of the industrialization process in Asia. *Source Handbook of Urban Studies*, SAGE Publication 2001, P421

Other studies also point out the host of problems brought about by the development of mega-city regions in East Asia, such as poor urban and rural environment, high construction cost and low operational efficiency. According to the Asian Development Bank's *Annual Report 1997*, "Without proper intervention to address these issues, the mega-city regions will become more crowded, environmentally polluted, unwholesome in production, expensive in costs and socially divided. And if foreign investment is disrupted, there will be a downward spiral: the comparative advantage of megacities will shrink, and the resources available to solve these growing problems will be reduced accordingly" [17]. This proves that the problems are prevalent across highly urbanized regions of the Asia–Pacific region.

3.3 Guidance of Spatial Planning on Urban Development

3.3.1 Origin of the Concepts of Spatial Planning and National Spatial Planning

Since the 1980s, the planning theory and practice of developed Western countries are attaching more importance to the integrity and coordination of spatial development. Returning to the planning system centering on physical space planning, they gave more value to economic, social, and environmental goals. And to avoid different names for urban–rural planning systems from country to country, the European Union named this kind of planning with integration and coordination functions and of strategic importance "Spatial Planning". Gradually, this type of planning become the collective term for planning systems at various levels in all countries.

National spatial planning refers to the overall comprehensive arrangement and strategic deployment for continuous coordination of the factors of economy, society, resources, and environment nationwide. It is a macro-scale spatial plan that is usually formulated by the central government or other high-level institutions. A long-term basic plan of the national spatial development framework for more than 10 years, it also serves as the planning of physical space that involves comprehensive allocation of major economic activities and resources on a broad scale. Strictly speaking, it originated from the Netherlands.

For most countries, national spatial planning is an indispensable means for the country to improve the market economy system, enhance competitiveness, and administer macro-control; it is a means for the central government to prevent and correct market failures and make government intervention under a completely free economic system. Politically, national spatial planning plays the role of embodying democracy, it is the expression of democratic ideology and popular participation. In terms of administrative system, national spatial planning is the general principle coordinating the interests of various departments of the state and between the central government and local governments, that is it has the role of administrative coordination.

3.3.2 Evolutionary History of the National Spatial Planning

In 1912, the British biologist P. Geddes put forward the theory of regional planning in his book *Cities in Evolution* for the first time. In the following century, regional planning was given importance and developed to fix the problems of urban and regional development arising during the rapid industrialization and urbanization. In 1906, the Belgian government formulated a public facilities investment plan for the Belgian Congo, which is considered the world's first national medium- and long-term plan in the true sense of it, marking the beginning of medium- and long-term planning in market economy countries [18]. The former Soviet Union carried out a comprehensive regional study in 1920, and initiated the national economic zoning in 1921, advocating organized and step-by-step regional development of the country under the guidance of the national plan. Other examples include the regional plan for the Newcastle coal mines of the United Kingdom in 1923, the influential New York regional plan of the United States in 1929, and the Tennessee Valley regional plan across seven states based on the watershed in 1933. These regional planning practices of different contents and at various levels laid the foundation for the great development of spatial planning after World War II.

3.3.2.1 Great Development of National Spatial Planning Brought About by Rapid Industrialization

In 1944, Abercrombie of the United Kingdom presided over the preparation of the Greater London area planning, which became the trailblazing attempt to carry out regional spatial planning centered on large cities. The significance of this practice is twofold: first, it signified that the study of London's development was expanding from the city to the region, and second, it provided technical support for the rapid recovery and development of the British economy after the war in terms of space. From a worldwide perspective, the major reasons for spatial planning at the macro level, especially at the national level, to be gaining more and real attention is that the country is at the stage of rapid economic development and the central government has a strong capacity to intervene. In the 1950s, Japan's economy post-war began to recover. To promote the development of regional economy and achieve balanced development across the country, the *National Spatial Planning Act* was enacted in 1950, which was followed by the "First National Comprehensive Development Plan" and a series of comprehensive development plans for large economic zones in 1962 [19]. As one of the Four Asian Tigers, South Korea also started its national planning in the 1970s when its economy was booming, and in 1972 it promulgated the "First Comprehensive Territotial Development Plan" to promote rapid economic development by cultivating new centers of economic growth. The Netherlands was the first country to propose national spatial planning; it issued the first national spatial planning report in 1960, putting forward the goal of balancing fairness and efficiency in the rapid development of the national economy after World War II. Since then, similar planning was carried out in France, Germany, and other countries.

3.3.2.2 More Importance Was Attached to National Spatial Planning Due to Population Growth and the Pressure of Energy and Environment

Since the 1970s, with the drastic growth of the world population and such events as the oil crisis, industries and population continued to pour into large cities, which resulted in the deterioration of urban environment. Given the circumstances, many countries realized the importance of conducting holistic regional research, and the depth and breadth of spatial planning have thus been greatly strengthened. Some Asian countries, such as Japan, South Korea, and Singapore, have placed greater emphasis on regional spatial planning due to their large populations and resource scarcity. In its "Third National Comprehensive Development Plan" promulgated in 1977, Japan adjusted the country's development plan in response to energy constraints on development. South Korea clearly stated the goal of protecting natural environment of the country in its "Second Comprehensive Land Development Plan" promulgated in 1982. The third national spatial planning report of the Netherlands in 1973 set "selective economic growth" as the goal, making reducing environmental pollution one of the strategic goals of national development. France and Germany divided their countries into interconnected regions according to the needs of economic development and made planning respectively. In the UK, regional economic planning committees were established to coordinate space-use planning (the charge of the Department for Environment, Transport and Regions) and economic planning (the charge of the Department for Trade and Industry).

3.3.2.3 Improving National Competitiveness in the Era of Globalization Become an Important Driver for the Resurgence of National Spatial Planning

Since the 1990s, the dramatic progress of economic globalization, trade liberalization and networking has intensified the competition between countries and regions. To attain a higher position in the global competition system, local governments put more emphasis on regional alliance to form synergies for global competition. Almost all countries and regions are involved in regional economic organizations to varying degrees, which brings about the demand on various aspects including overall regional development, coordinated development of urban and rural areas, coexist with sound ecological environment, and co-construction and sharing of facilities. The goal of planning is no longer limited to solving specific problems within one region, but to enhance the competitiveness of the region itself to obtain more development opportunities. Therefore, large-scale regional planning or inter-administrative region planning have been attached more importance and attention, especially so for national spatial planning. For instance, the national spatial planning of Japan, South Korea, the Netherlands, etc., around 2000, as well as the space policies of the United Kingdom and Germany have all clearly stated the goals of improving national competitiveness for global competition and going all out to build internationally competitive urban areas. Japan has proposed to build the Tokyo metropolitan area into a world city, and the Netherlands stated to make the Randstad area an important economic center of the Europe that is comparable to the Greater Paris and Greater London. Regional development plans covering different countries or continents are also developing rapidly, examples include the ESDP, the spatial planning of the eight Eastern European countries, and the planning devoted to the regions around the Andes in Latin America (Bolivia, Colombia, and Peru).

3.3.3 Main Models of National Spatial Planning

3.3.3.1 Three Types of Planning Systems

The emphasis, policy orientation and planning theory of spatial planning vary from country to country due to their different historical backgrounds, social systems, economic conditions, geographical environment, and cultural norms. But in general, spatial planning is employed by the national and local governments to guide and regulate regional economic development. According to French and German scholars, there are two different types of market economies in modern times, one is the "Rhine model" (developed capitalism in Western and Northern Europe), and the other is "Anglo-Saxon model" (British-American model). The "Rhine model" adopts medium- and long-term economic development plan, has state-owned economy of considerable scale, and values government's initiative in handling the relationship between the government and the market, i.e., government's intervention is important. The "Anglo-Saxon model" pursues a free market model with less government intervention in the market, and there are much less state-owned enterprises than in the "Rhine model" countries. The "Soviet model" pursued by socialist countries such as the former Soviet Union, Eastern Europe, Vietnam, North Korea, and China before the Reform and Opening Up is another important economic model (planned economy). This model believes in government intervention in the economy, puts state-owned economy at a dominant position, and all construction is implemented in strict compliance with plans. In summary, the political and economic system of the country determines the overall preparation and implementation model of spatial planning, which is not only a technical process, but a political one, an action on the part of both government and society. The spatial planning system can be categorized into the following three types based on existing theories and practices:

(1) System of Strong Government Intervention

This type is represented by socialist countries such as the former Soviet Union and Eastern Europe, as well as Western capitalist countries including France, the Netherlands, Greece, Japan, Singapore, etc. In these capitalist countries, there is generally a long tradition of centralization, collectivism and nationalism are their leading social values, land is owned by the state according to law or the state has a strong say in the development of private land, the central government and governments at all levels are in charge of planning, regional governments are often set up to coordinate the development of urban clusters in a larger area, that is, a two-tier administrative leadership was established and planning is one of the main responsibilities of governments at all levels, and sound planning institutions and mechanisms are in place. Meanwhile, due to the small size of these countries, there is generally an inherent need for intensive development of regional space.

The former Soviet Union is a typical example of national spatial planning under the planned economy. The "All-Russian Electrification Program" developed in 1920 under the leadership of Lenin was the world's first nationwide long-term plan for the national economy. Based on the theory of rational organization of production regions, the program, using electrification as the driving force, divided the then European USSR into eight regions and stated the characteristics of each region and the specific tasks of restoring construction. In the 1920s and 1930s, comprehensive plans were made for the oil drilling in Baku and the power industry in Dnepro, marking the beginning of regional planning. Since then, the economic zoning with the aim of promoting and coordinating planned, proportional, and balanced development of the economies of zones at different levels has become the routine task of the State Planning Commission of the former Soviet Union. In the 1920s, under the guidance of the idea that "the basic economic zone, economically perfect as much as possible yet not self-sufficient, is a unique part of the national economy", the country was divided into 21 basic economic zones (including 12 in Europe and 9 in Asia), and it was planned to further divide them into 140–150 secondary economic zones and about 3000 grassroots economic zones on this basis. By the end of the 1930s, the basic economic zones were merged into 13. Up to the eve of World War II, the three-tier economic zoning was a key regional unit for formulating medium- and long-term economic development plans for the entire former Soviet Union and the union republics, as well as the first administrative regions of the Krai, oblasts, and autonomous republics, and played an important role in the construction of industry, energy, transportation, and urban layout. From the *Third Five-Year Plan* in 1937 on, the spatial planning at the national level was formally formulated. Up to the collapse of the former Soviet Union in 1988, the country was divided into three major economic zones, 149 economic administrative regions, and 3225 intra-oblast economic zones.

Before their disintegration, the socialist countries of Eastern Europe followed suit with the Soviet Union in their economic management system, that is they all regulated the national territory and regional development through national economic planning. They upheld the theory of regional division of labor and balanced distribution of productive forces in promoting the rapid economic development and productivity layout of the whole country and major regions. For example, the Hungarian development plan is jointly prepared by the Planning Commission and the former Ministry of Construction and served as the strategic decision for the long-term development of the country. The plan, officially completed in 1960 and approved by the Hungarian Parliament in 1971, basically achieved its goals after a long period of practice[6].

France is a highly centralized state with a history of government intervention. From the central government to governments of all levels, planning institutions of strict top-to-bottom hierarchy have been established. Since 1959, the country has systematically set up several syndicate-style joint committees among 300-odd municipalities across the country, and the government has established Regional Research Institute (RPI) in the main urban areas that are subject to audit. Therefore, the master planning concept of the central government and regions can basically be implemented tier by tier, and it was possible to formulate and implement such plans as the "balanced metropolis" on a national scale (1965) and the axial development plan for the densely populated urban space centering in Paris (1965), which was revised twice in 1979 and 1980, respectively.

The Netherlands is known as a "Planned Country", and its urban planning is not limited to cities, but treats the entire country as a city. The country formulated the "National Planning Report" five times in 1960, 1966, 1976, 1990 and 2000, respectively. Compared to other European countries, the spatial planning system in the Netherlands is both thorough and detailed. Its national planning focuses on the general overview of the ideal spatial organization for the future, and it pays attention to urbanization-related issues. Structural planning at the provincial level plays an important role between national planning and urban land use planning. Construction activities can only be carried out with the building permit which was granted only when it complies with the urban planning.

Japan's land resources planning system is divided into four levels: (1) national comprehensive development planning, (2) construction planning of the three metropolitan coordinating regions, development planning of the seven major areas, and planning of special areas (such as islands, mountain villages, and underdeveloped areas), (3) (long-term) comprehensive development planning of Tokyo Metropolis, Hokkaido, Osaka and Kyoto Prefectures and other prefectures, and (4) (long-term) comprehensive development planning of towns and townships. According to the *National Spatial Planning Act* (1950) and relevant supporting regulations, the Land Resources Agency (merged with the former Ministry of Construction, the Ministry of Transport, and the Hokkaido Development Agency in 2001 to form the Ministry of Land, Infrastructure, Transport and Tourism) is responsible for formulating national planning, planning of major regions, and planning of special areas. Tokyo Metropolis, Hokkaido, Osaka and Kyoto Prefectures and other prefectures formulate their own long-term comprehensive development plans by considering the suggestions of towns and townships, and coordinate development plans within its own region and with the surrounding regions. The national plan formulated by the central government

[6] Quoted from *Theory and Method of Land Resources Planning*, Wu Cifang. Beijing: Science Press, 2005. 65–66.

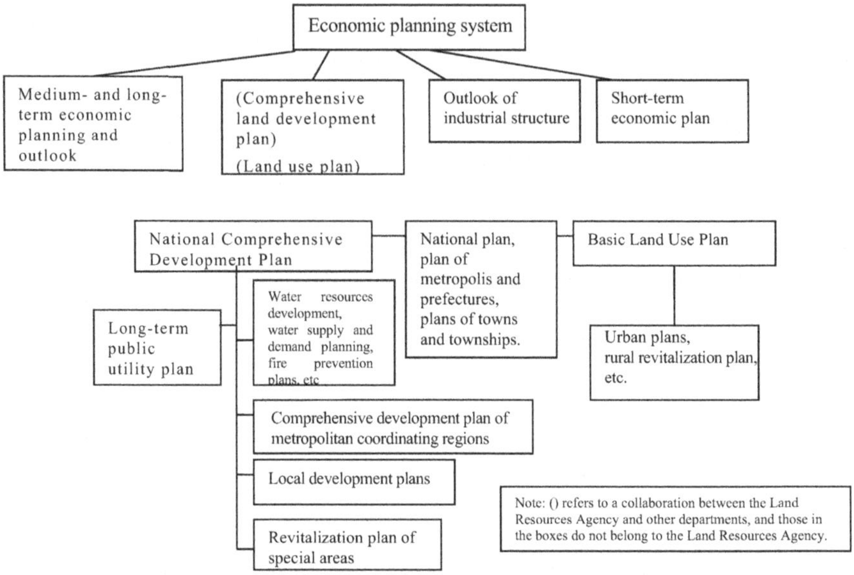

Fig. 3.3 Land resources planning system of Japan

is approved by the cabinet meeting after being deliberated by the Land Resources Planning Council. Various departments of the state formulate 14 long-term plans (5–7 years) for public investment and construction in line with the national plan, making public investment in areas that are the charge of the state and subsidizing local governments (Fig. 3.3).

(2) System of Full Local Autonomy

The United States is the representative of this type. Although the country does not exercise centralized and unified management of planning, and the various planning are prepared by regions or cities themselves, it guides the allocation of space resources across the country through legislative and financial means.

As a federal system country, the central government of the United States has basically no say in planning, the power of which is delegated to the state governments who further delegate the power to the self-governing institutions at various levels below the state level. The power and influence of the federal government in domestic affairs is mainly realized through the allocation and direction of federal funds. The federal government has little to do with regional planning, nor does it have any master plan that governs the planning formulated by state and local governments. The classic example of national spatial planning in the United States is the Unified Development of the Tennessee River System Plan in the 1930s and 40 s that spanned 7 states in the southern heartland of the country, which gave birth to the Tennessee Valley Authority (TVA). The experience was learned and emulated by many countries later, but no similar planning was ever carried out in the US since then.

Although there is no clear regional planning in the United States, there is no lack of cross-state "regional planning" documented in some relevant files. Taking the planning of national large-scale public works (such as transportation) for instance, state governments were subsidized based on planning, and laws and regulations were formulated to manage the development and utilization of resources, which amounted to indirect intervention in state planning.

In recent years, the U.S. government has been making conscious effort to strengthen government regulation and proposed the concept of "growth management". It attached importance to resource sharing, complementary functions, and shared obligations among administrative areas in the region. To curb the unchecked disorderly development of regions, the government has also encouraged the establishment of various inter-administrative area joint mechanisms for coordinated administration and enhanced the government's regulatory capacity through the allocation of federal funds. For example, it is stipulated that local regions can only be granted the federal highway funds if they have a comprehensive plan in place.

The EU of 25 member states is the regional political and economic bloc of the highest degree of integration in the world today. And with the continuous progress of the EU integration, its member states have arrived at broad consensus on the establishment of a guiding framework for spatial development at the union level. At the informal meeting of EU Member States' Ministers for Spatial Planning and Regional Development in 1994, the 15 member states agreed on the principles for drafting the "European Spatial Development Perspective" (ESDP), and in 1999 published the official document of the ESDP to guide the spatial development of the member states. Although ESDP only serves as a guiding document rather than a directive, the EU requires its members to follow the principles and policies of the ESDP in their development plans and facilitates their implementation through "structural funds". The guiding principles and policy objectives of the ESDP reflect the common values concerning spatial development and planning across EU countries, and the choice of policy takes into consideration the diversity of countries and regions.

It took years of the EU countries to complete the ESDP, a planning policy of spatial integration, which is an important part of EU regional policy. In terms of both its formulation method and implementation mechanism, ESDP fully reflects the power of policy of different countries and regions to use planning to achieve common goals and promote the development of their respective countries and regions.

(3) System of Moderate Intervention

The representatives of this model include the United Kingdom, Germany, Denmark, and Italy. The central government made conscious, moderate intervention and overall coordination, and planning is carried out level by level according to a unified procedure under one competent authority. The central government has guiding and adjudicating power over local planning to a certain degree, and exercises regulation in various aspects including laws, policies, and economy.

The United Kingdom established the Joint Town Advisory Committee in Scotland as early as 1928 and went on carrying out regional planning on a national scale. Although the practice of regional planning after WWII mainly concerned itself with economic planning, it served as a kind of urban spatial planning from the very beginning as it took full consideration of such spatial elements as infrastructure, new industrial areas, residential areas, ports, airports, etc. The urban planning system mainly includes structure plan, local plan, and unitary development plan. Structural plans are prepared by county planning departments and become effective upon approval by the Minister of the Environment of the central government. As for metropolitan areas, comprehensive development plans are compulsory. At the national level, the Ministry of Environment is responsible for preparing the Planning Policy Guidance (PPG) and the Regional Policy Guidance (RPG) to guide the development planning of various regions and cities. In the new round of discussion about the reform of the urban–rural planning system, regional planning covering a larger range was proposed, and it was required to use this as the unit to prepare regional development plans.

Germany is a federal State, but there are close ties between the central, state, and municipal governments. The Federal Ministry for Regional Planning, Architecture and Urban Development of the country coordinates development between the federation and the states through the preparation of comprehensive regional plans, and there is also a specialized body, the Ministerial Conference for Spatial Planning (MCSP)[7]. For spatial planning in Germany, each state is responsible for formulating and implementing its own land resources planning under the guidance of the basic law of the Federal Government. According to the basic law—the *Federal Space Development Act*, the Federation stipulates the concepts, principles, and procedures for the development of space throughout the country, and the states have absolute autonomy and legislative power to formulate its own laws and plans by considering the concepts and principles of the *Federal Law*. The federation would also work with the states to develop policies and basic guidelines that affect the federation. The basic law aims to promote the development, construction and protection of the whole country and regions, coordinate socio-economic goals with ecological goals, and maintain balanced and sustainable development of regions on a large scale through comprehensive and higher-level planning. The national planning system is divided into four levels: (1) outline of the federal spatial development policy, (2) state development plan, (3) regional planning, and (4) municipal, town, and village planning. The German government usually guides and regulates the formulation and implementation of development policies by providing infrastructure and financial support to the regions (Fig. 3.4).

In addition, the "regions" in the Italian administrative system, with considerable administrative power, are the key bodies for formulating regional plans and implementing industrial development, and their main responsibilities are to formulate regulations and make arrangement for major construction projects. Regional planning in Denmark is formulated at the following four levels in accordance with the law: national and regional planning regulations, regional planning regulations for metropolitan areas, municipal planning regulations, and urban and rural zoning regulations, and is administered centrally by the Ministry of Environment.

[7] The body is advisory in nature and responsible for coordination.

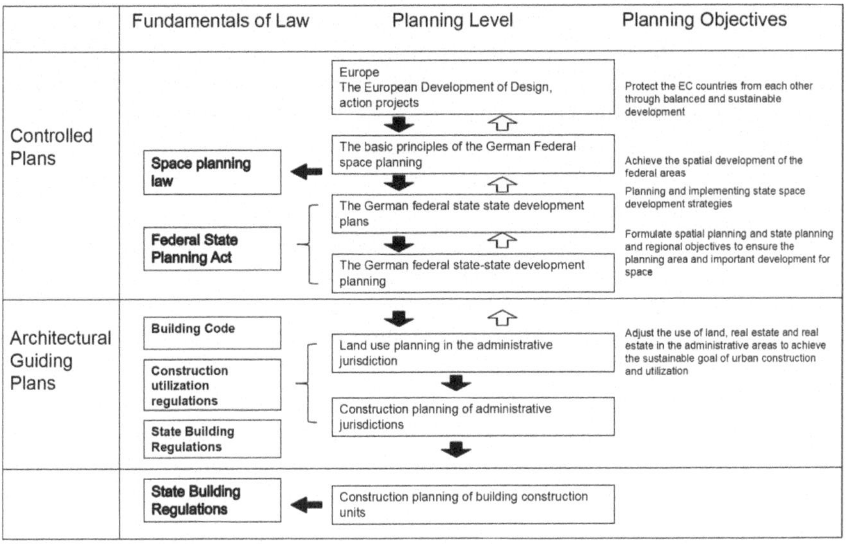

Fig. 3.4 Spatial planning system of Germany

3.3.3.2 Major Models of Planning

(1) Japanese Model Centered on Guiding Development

Japan formulated a total of five National Comprehensive Development Plans since 1962, using the opportunities offered by the plans to fix the problems of population and industrial development brought about by industrialization and urbanization at the national level comprehensively. The five plans have played a positive role in the efficient and sound use of space resources nationwide and made an important impact worldwide.

Enacted in 1962, the first National Comprehensive Development Plan (1st Comprehensive Plan) proposed the planning objective of achieving balanced development between regions in response to the following problems occurring during the rapid economic growth of the country in the 1950s: excessive concentration of enterprises in the coast, continuous expansion of large cities, widening of regional disparity, and excessive use of land and water. The Concept of Industrial Belt Layout along the Pacific Coast and the Income Doubling Program were introduced to solve the above problems in a comprehensive manner. In terms of spatial policy, the idea of developing "bases" to promote the development of local economy were put forward, that is, apart from Tokyo, Osaka, and Nagoya, to focus on the fostering of two types of "bases" centered on industrial development and urban functions.

Enacted in 1969, the second National Comprehensive Development Plan (2nd Comprehensive Plan) put forward the objective of developing and utilizing national land resources on a larger scale through large development projects such as industry, animal husbandry, and transportation in response to the following phenomena emerging in the 1960s: worsening of regional disparity, continued concentration of enterprises to the coast, and surplus of rural labor. In terms of spatial policy, measures such as adjusting the functions of large cities, ridding them of industries, and strengthening the function of central administration were proposed. In underdeveloped areas, it was proposed to increase public investment, promote industrial development and tourism development, improve living conditions, and more importantly, put forward planning proposals for the construction of transportation networks. Considering the general industrial layout, the Tohoku zone, the central zone, and the southwest zone were proposed, as well as the concept of the seven "broader circles of living"[8] including the Hokkaido circle, the Tohoku circle, the capital circle, the central circle, the Kinki circle, the Shikoku circle, and the Kyushu circle. The 2nd Comprehensive Plan played an important role in Japan's economic development by extending the sphere of "bases" through large-scale development projects.

The third National Comprehensive Development Plan (3rd Comprehensive Plan) was enacted in 1977. In response to the environmental hazards and uneven density of regions caused by rapid industrialization, and resource shortage brought about by the oil crises in the 1970s, the plan put forward the comprehensive environmental objective of building a livable environment for people. Specifically, it aimed at using its limited land resources to cultivate a wholesome, culture-loving, and humane environment characterized by harmonious relations between nature and man by drawing inspiration from its history and traditional culture and giving full play to the strengths of the regions. Instead of sticking to the previous model prioritizing economic development, the spatial policy proposed to focus on the construction of living facilities of residence, education and culture, medical care, etc., by using "settlement circle" as the basic unit, curbing the concentration of large cities, and promoting balanced development. The plan also put forward the idea of technology-intensive cities in response to the strategy of building the country with science and technology. The 3rd Comprehensive Plan represented a major adjustment to the development concept as it attached importance to comprehensive environment for human settlements in its choice of goals, showed more respect to local authorities in their own administration, and reflected on putting economic benefits above all else.

The fourth National Comprehensive Development Plan (4th Comprehensive Plan) was enacted in 1987. In response to the challenges of deepening polarization of Tokyo, changes in industrial structure, mounting employment pressure, aging society and internationalization in the 1980s, the plan proposed to establish a decentralized land development structure of multi-axis and multi-polar, to prevent excessive concentration of population, economy and administrative functions in any single region, be fairer and more efficient, strengthen the connection between the central and local governments, promote the complementary relations and integration between regions and the world, and build a world city. The 4th Comprehensive Plan played an important role in distributing Tokyo's overly concentrated urban functions and

[8] "Broader circle of living" is the basic unit of land and regional development. With a radius of 30–50 km, it is larger than the commuting circle, shopping circle, and school district circle, and has a central city equipped with urban functions.

made a positive impact on the formation of diverse and distinctive areas, especially in the aspects of construction people, materials, and information networks across the country.

The fifth National Comprehensive Development Plan (5th Comprehensive Plan) was enacted in 1998. In response to the challenges of globalization and informatization, relocation of traditional industrial zones to the third world, and the decline of total population in the 1990s, the plan came up with the following measures in compliance with the trend of internationalization and informatization: forming an international circle of exchange of wider range, establishing a multi-axis territorial structure that is open to the world,[9] letting its people enjoy the beautiful nature, and urban regeneration. In terms of development model, the following important policies were proposed: government-enterprise cooperation, regional collaboration between small and medium-sized cities and surrounding rural and mountain villages, collaboration between central cities of different levels, promotion of international exchanges at the regional level, and regional cooperation aimed at promoting the development of industry, welfare, education, culture, and tourism. The 5th Comprehensive Plan embraced the planning ideas of economic globalization and sustainable development; replaced the "one pole, one axis" structure with a multi-axis spatial structure; for urban system, the traditional hierarchical structure was replaced by the flat network structure; self-reliance of and complementary relations between cities were encouraged to cut their dependence on Tokyo; and in utilities such as transportation and communication, high efficiency, equal accessibility and minimized impact on environment were aimed for.

Analysis of the five comprehensive plans showed the most distinctive character of their formulation, i.e., it invariably aimed at solving the problems in regional development by taking account of the changing international situation and the needs of national development. It was particularly so in spatial policy, that is to promote the goal of balanced development through the spatial organization of industries and cities and towns, as evidenced by the "bases", "broader circles", and "settlement circles" of early stage and "multi-axis, multi-polar" later. Over the past 50-odd years, the five comprehensive plans have become the most important guiding principles for Japan's economic development, land resources development, resource protection and balanced regional development, contributing to the country's socioeconomic development.

(2) Korean Model Centered on Controlling Imbalance

South Korea conducted a total of four land resources planning since the 1970s in response to the legions of problems arising from its rapid industrialization and urbanization.

The first Comprehensive Territorial Development Plan, 1972–1981 (1st Comprehensive Plan) put forward the general objectives of maintaining sustained economic growth, improving land use efficiency, and improving the living environment of the

[9] They are the West Japan Land Axis, the Tohoku Land Axis, the Sea-of-Japan Land Axis, and the Pacific New Land Axis.

people in view of the rapid economic development since the 1960s and its focus of investment being on the highly productive industrial sectors, fostering urbanization centered on Seoul and Busan. Its spatial policy during this period proposed to build large-scale industrial infrastructure, establish transportation, communications, water resources and energy supply networks, strengthen regional functions in backward areas, and develop industries compatible with regional features. Although the 1st Comprehensive Plan contributed to the rapid improvement of the country's national strength and industrialization to a certain extent, it had little effect on solving the problem of greater concentration of population and industry in the capital circle centered on Seoul (Table 3.2).

The second Comprehensive Territorial Development Plan, 1982–1991 (2nd Comprehensive Plan) was developed to fix the problems of greater concentration in the capital circle and geographical imbalance. The plan proposed to induce the population to move and settle in the local areas, establish a multi-core structure of the country, alleviate the polarization of Seoul and Busan, foster local cities with great potential into new growth points, curb and regulate the growth of the two major cities of Seoul and Busan, strengthen transportation and communication planning, heighten the development of backward areas, and formulate plans for developing backward areas with development potential such as abundant in tourism resources and regional specialties into special areas, and provide active support for their development. The 2nd Comprehensive Plan succeeded in maintaining sustained economic growth in a larger region and improving the living environment of its people.

The third Comprehensive Territorial Development Plan, 1992–2001 (3rd Comprehensive Plan) was still the result of the over-polarization of Seoul-Busan. The plan proposed to construct a decentralized land pattern, shifting from the negative balanced development method that curbs the concentration in the capital circle to an active mode featuring local development centers. It further proposed the following: set up business groups in large cities; in small and medium-sized cities, select and cultivate major industries that are compatible with the characteristics of the cities; form new industrial zones in the central and western regions of the country; induce exchanges and competition with China; disperse the industrial functions of the centralized Gyeongbu axis to the west coast; build a comprehensive high-speed exchange network, expand domestic and foreign trunk line exchanges, and establish an interconnection system between transportation, circulation, and communication facilities; and take more effective measures to implement the territorial resources planning and clarify the responsibilities of the central government, local governments and the private sector. The 3rd Comprehensive Plan laid the foundation for Korea's economic growth and greatly improved the living environment of the people. The concentration of population and industries in the Seoul-Busan axis and the capital circle continued to intensify, and by 2000, 46.5% of the population, 88% of large enterprises, and 84% of the national public institutions were in the capital circle whose land only accounts for 11.6% of the country's total.

The fourth Comprehensive Territorial Development Plan, 2000–2020 (4th Comprehensive Plan) was created for the sake of overhauling the unbalanced spatial development pattern with new ideas. It put forward the objective of forming a richer

Table 3.2 Main contents of Korea's four comprehensive territorial development plans

	1st Comprehensive territorial development plan, 1972–1981	2nd Comprehensive territorial development plan, 1982–1991	3rd Comprehensive territorial Development Plan, 1992–2001	4th Comprehensive territorial development plan, 2000–2020
Basic objectives	Efficient use & management of national territorial resources; grow the base for development; resource development & conservation of natural resources; improve the living environment of the people	Dispersion of population to local areas; nationwide development of territorial resources; raise the level of national welfare; protect the natural environment of the country	Build a decentralized territorial pattern; establish a productive and intensive land use system; improve national welfare and protect the environment Lay the foundation for the reunification of the Korean Peninsula	Balanced territory; green territory; open territory; and unified territory
Basic strategies	Build large-scale industrial bases; grow transport, communication, and energy supply systems; strengthen regional functions in backward areas	Establish a multi-core structure of the national territory; wide-area development; alleviating geographical disparities; development of backward areas	Centralized control of the capital circle and local urban development; cultivation and industrial sophistication of the southwest industrial belt; construction of a highway network; strengthen the execution of territorial resources planning; development and management of inter-Korean exchange zone	Construction of open territorial axes; enhance competitiveness of local regions; environment-friendly territory management; construction of high-speed transportation network and information network; lay the foundation for inter-Korean exchanges and cooperation

(continued)

Table 3.2 (continued)

	1st Comprehensive territorial development plan, 1972–1981	2nd Comprehensive territorial development plan, 1982–1991	3rd Comprehensive territorial Development Plan, 1992–2001	4th Comprehensive territorial development plan, 2000–2020
Projects Planned	Construction of industrial infrastructure; expansion of transport & communication networks; urban development; water resources development	Settlement system and population relocation; improvement of the national living environment; expansion of the bases for national territory development; land use management	Local urban development and centralized control of the capital circle; industrial layout and tourism development; construction of a comprehensive transportation network; territory and resources management	Regional economic development; development of tourism and cultural industry; construction of high-speed transportation network and information network; construction of international passengers & logistics base
Measures	Development of bases for growth; development of river basin circles	Development of decentralized bases and living circles	Multi-polar, multi-core pattern of development; development of regional economic circles	Development of broader circles and territorial development axes
Circles/ Regions Involved	4 major river basin circles (Han River Circle, Geumgang Circle, Nak-Dong River Circle, Yeongsan River Circle), 8 middle circles, 17 small circles	5 living circles of large cities, 17 living circles local cities, 6 living circles in rural areas	Not clearly defined	10 major broader circles, capital circle, 2 major development axes, 6 small development axes

Source Theory and Method of Territorial Resources Planning, Wu Cifang et al. Beijing: Science Press, sorted out in 2003

pattern featuring "balanced territory, green territory, open territory, and unified territory", aiming at the coordinated integration of North and South Korea. In terms of spatial policy, it was proposed to build an open and unified territorial axis plan, building three development axes in the interior of Korea and three coastal territorial axes in the coastal zones, to improve the level of development in the hinterland and the competitiveness of coastal regions. It was further suggested to cultivate a wholesome and sound environment, turning mountainous areas into beautiful pastoral living places and cultural and leisure spaces through restrictions in height and floor area ratio, promote urban greening, protect wetlands on the west coast, and build a high-speed transportation network and a land intelligence network to make every place of the country easily accessible.

In the 4th Comprehensive Plan, South Korea, in response to the strong inertia of market forces, managed to improve the quality of the living environment through guidance of industrial development, cultivate the development capacity of the local region itself, and promote the coordinated development of the region. Of all the measures it took, the fundamental improvement of infrastructure conditions played an important role in the balanced national spatial pattern.

(3) Dutch Model Centered on Strengthening Competitiveness

In 1960, the Dutch government formulated its very first report of national spatial planning. The report proposed to take the objective of balancing fairness and efficiency as the starting point for national spatial planning. Based on this principle, the report went on pointing out that on one hand, it is imperative to control the distribution of the population throughout the country, dispersing some of the population and jobs in the Randstad region, and making sure various social and cultural facilities are evenly distributed throughout the country to compensate for the disadvantages of remote areas. On the other, it is necessary to gather certain important economic functions in Randstad, such as transnational corporations, strategic decision-making centers, ports and port industries, export industries, etc., to improve the development capacity of the central area.

In 1966, the Dutch government prepared the second report of national spatial planning. In response to the rapid development of private cars and the trend of suburbanization of cities at the time, the report proposed to spread the population and jobs of Randstad outward along the convenient transportation routes, that is to distribute the population in groups in urban agglomerations according to the principle of "controlled decentralization". This way not only was the historical landscape of Randstad kept intact, but the living environment of its inhabitants was also improved as well. In addition, to make the Netherlands part of the large urban agglomeration of Western Europe, the report also proposed the development of Randstad's urban wings of the south and north as part of the central European urban system.

In 1973, the Dutch government began working on the third report of national spatial planning. In view of the trend of decreasing population in some major cities in the Randstad area in the 1970s, as well as the gradual migration of many urban rich people to small towns and even rural areas, the concept of "selective economic growth" was proposed to give special priority to existing towns, controlling the dispersion of population, and increasing current urban population to prevent cities from shrinking. Meanwhile, "urban area"[10] was treated as an important concept of the country's national spatial planning; the report believed that Randstad is a complex of many urban areas, and public transport systems will be developed within

[10] The so-called "urban area" refers to the whole of the central city and its surrounding growth centers (new cities) organically connected by transportation.

and between urban areas, this way, the development capacity of regions will be strengthened through the improvement of infrastructure.

In 1988, the Dutch government published its fourth national spatial planning report. Compared with the third report, this report had three distinctive characteristics. First, it placed special emphasis on the improvement of the quality of daily life environment and of spatial structure and took "sustainable development" as one of its basic starting points. Second, the significance of Randstad was restored to improve the country's international standing and enhance its international competitiveness. Third, great importance was attached to the cooperation between central, provincial, and local governments, as well as between the public and private sectors. The report also proposed to gradually expand the economic core area from Randstad to form a central Dutch metropolitan coordinating region of cities, the center of which is an agricultural area more expansive than the Randstad's green heart. In this coordinating region, Rotterdam is the world's largest port, Amsterdam, the capital of the Netherlands, is the financial center and an important airport in Europe, and The Hague is the seat of the Dutch government and many international institutions. In future, the region will be an internationally competitive urban area, comparable with such large cities as Paris, London, Brussels and so on.

In 2000, the Dutch government published its fifth draft policy document concerning national spatial planning. The plan put forward visions and forecasts of various land use trends in the Netherlands including housing, employment, infrastructure, entertainment, sports, natural environment, vacant land, agriculture, and water, and kept the two main objectives of improving the quality of space and guiding the use of space by economic and social activities. To articulate the objectives more clearly, the fifth planning adopted an analytic hierarchy which categorized the country's space into three tiers: the base tier, the network tier, and the application tier. The base tier refers to the natural material and ecological conditions on which spatial changes occur. The network tier refers to all roads, railways, waterways, pipelines and sewers, ports, airports, transfer stations and digital networks, which are further divided into the infrastructure network and transportation network. The application tier refers to the spaces of people's life, work, and leisure. Given the huge difference between urban and rural areas, and between developed and underdeveloped areas at the application tier, the plan stressed to maintain the spatial differences between cities and villages, between different cities and different villages, provided that social fairness is guaranteed, and to improve the quality of space. The plan also expounded the objectives and constructed relevant policy frameworks in terms of international cooperation, urban and rural areas, urban networks, and water resources. To be more specific, the key goals include strengthening the competitiveness of the Netherlands, focusing on the construction of Rotterdam Port and Amsterdam International Airport, introducing the concept of "Delta Metropolitan Region", and building the Randstad metropolitan coordinating region into an international city network. In urban and rural planning, three strategies for government intervention were proposed: first, intensive land use for urban areas; second, comprehensive use of land for rural areas; and third, transform urban and rural spaces to better meet demand. The plan went on dividing the land of the Netherlands into three categories: red-line areas (built-up areas), green-line areas (ecological areas), and transition zones. And detailed spatial objectives will be determined, and different strategies and policy measures of spatial intervention will be adopted according to the characteristics of each area/region.

The spatial planning of the Netherlands is more systematic and mature in theory and practice as it always: attached equal importance to the balanced development of the region and the enhancing of the competitiveness of the core areas, analyzed the development of the core area from the perspectives of the Europe and the world at large, and treated the urban–rural relationship in its analysis. So, there is a lot we can learn from the technical methods it adopted including the series of scientific analysis methods, policy guidelines by regions, and control measures.

(4) German Model Centered on Coordinated Regional Development

With the German unification, the formation of the European Common Market, and the concept of sustainable development taking root, the German Ministry for Building has formulated three important documents since the 1990s: the Report of Spatial Order Planning (1993), the Policy Orientation Framework of Spatial Order Planning (1993), and the Policy Measures Framework of Spatial Order Planning (1995). Of the three, the Report of Spatial Order Planning is the first comprehensive spatial development planning report of the federal government since the country's unification.

The main content of the Report of Spatial Order was to formulate regional policies and measures to promote economic development, balanced regional development and job creation; to establish the objectives, principles and relevant provisions concerning regional land improvement, such as the minimum standards for economic development, population movement, land-use structure, environmental protection and infrastructure of each region; and to put forward the general requirements for the construction of large-scale national infrastructure projects.

The policy framework of spatial planning at the federal level treated "region" as the level to take action to implement the spatial order; stressed on thinking about spatial development in a European context, and on cooperation with the EU when considering the development of large cities and border towns; emphasized the particularly important role of transport and environment in national space development; further clarified the rights and duties of the competent authorities of the federal government and local governments in the preparation and implementation of spatial order planning and state-wide planning.

The German Ministry for Building published nine major development objectives and tasks in 1996 as the task of German spatial planning in the twenty-first century: (1) make more efforts to disperse the population of settlements throughout the country; (2) establish a network of urban systems; (3) grow the connection between cities and the surrounding areas; (4) protect the ethnic and cultural diversity of agricultural areas; (5) mix living, working, and recreational land according to sound arrangement; (6) ensure the sustainability of the basic natural conditions; (7) build affordable transportation; (8) cooperation with its European neighbors; and (9) the construction of capital region. The central ideas of German spatial planning in the twenty-first century was hinted by these objectives: distributed centralization, networking of urban structure, and urban–rural integration. A useful exploration of the construction planning of the metropolis and its surrounding urban agglomerations is the overall spatial development plan combining Berlin with Brandenburg where it is seated and the regional space in its vicinity.

It can be seen from Germany's national space policy that coordinated regional development is the basis for the healthy development of the country, meanwhile, maintaining local diversity amid general development is also important in the era of globalization, and providing efficient infrastructure is crucial for comprehensive regional development.

(5) EU Model Aimed at Integrated Coordinated Economic, Social and Environmental Development

The ESDP is an important step in the European integration process. To achieve economic and social integration of the continental Europe, protect natural resources and cultural heritage, and attain a more balanced regional competitiveness, the 15 member states signed the report of spatial planning in 1999. With the expansion of the EU, the studies of sub-regional planning were later conducted based on this report.

As far as the EU is concerned, there is a serious imbalance in the economic potential within the region, which hinders coordinated regional development and sustainable development. While the gap in economic power between "prosperous" and "poor" regions is slowly narrowing, the gap between regions within most member countries is widening. To address this problem, the ESDP sets out a more integrated development objective: to achieve balanced and sustainable development through greater economic and social integration. The three guidelines for realizing the ESDP are: to develop a balanced and multi-centered urban system and a new urban–rural relationship; to ensure equal access to infrastructure and knowledge; and to achieve sustainable development, sound management, and the protection of natural and cultural heritage (Fig. 3.5).

The basic actions to achieve the ESDP are the Community competition policy; Trans-European Networks (TEN); Structural Funds; Common Agricultural Policy (CAP); environmental policy; research, technology, and development (RTD), and lending of the European Development Bank.

Determination of the Community Competition Policy The competition policy provides a series of guidelines at the community level to avoid such market abuses as cartels and monopolies. It controls mergers and acquisitions of companies and provides a framework for government subsidies, which have an impact on the geographical distribution of economic activities and the way trade is conducted across the EU.

Construction of TEN The Community is called upon to contribute to the organization and development of an infrastructure network throughout Europe, including transport, telecommunications, and energy supply. Build a well-functioned and sustainable transport system, particularly to help link the isolated, closed, and marginal areas with central areas. Through the construction of high-speed rail, major

Fig. 3.5 Illustration of the ESDP objectives. *Source* ESDP, European Commission 1999

European metropolitan areas are connected. And put in active effort to develop modern communication technologies and energy networks to support rural and inaccessible areas.

Construction of Multi-centered Spatial Structure The core of the European Union is a pentagon consisting of London, Paris, Milan, Munich and Hamburg and this region offers extensive global economic functions and services. The ESDP emphasized multi-centered, balanced urban systems and rural–urban partnerships in its spatial policy orientation, and aimed at building several dynamic, well-laid out international economic integration zones within the European Union that are linked by transport networks with access to internationally urbanized areas and the hinterlands under their influence (towns and rural areas of all sizes).

Implementation of Planning Through Structural Funds In addition to conventional subsidies, the conditions for applying for structural funds are determined based on the level of GDP per capita in smaller regions. Regional economies are promoted through the provision of funds. In larger areas of transnational-cooperation, emphasis is placed on the coordination of factors affecting the development of regional space (Fig. 3.6).

In terms of implementation, the ESDP provides an overall framework for policy implementation which is not the sole responsibility of any single authority but are the responsibilities of a range of competent departments of spatial development (land use, regional planning, urban planning) and sectoral planning (Fig. 3.7).

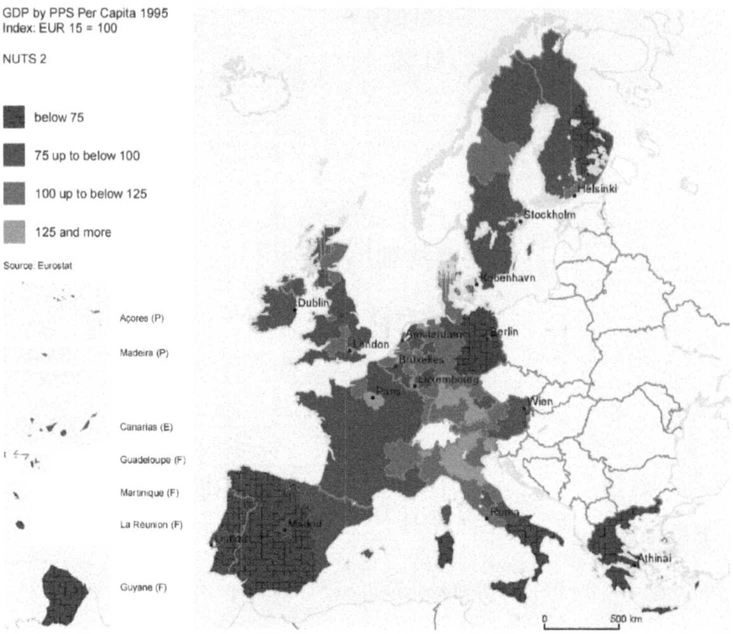

Fig. 3.6 Scope of structural funds. *Source* ESDP, European Commission 1999

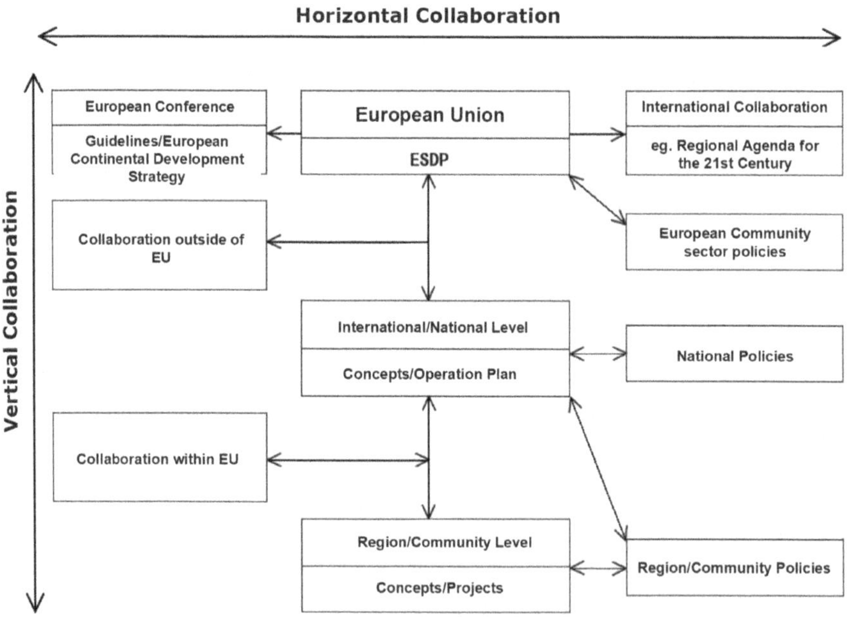

Fig. 3.7 Implementation framework of ESDP. *Source* ESDP, Joint Commission of the European Union 1999

3.3.4 Implementation Mechanisms of National Spatial Planning

3.3.4.1 Legal Means

In market economy countries, the law-based formulation and implementation of planning is an essential feature of those who are relatively successful in their spatial planning. The law serves as the basis for the formulation of and the guarantee for the implementation of the plans. Most developed countries have also had in place the basic laws that include such contents as the basic tasks, organization, and management of spatial planning, and formulated supporting laws, regulations, and policies in the process of implementation.

Japan enacted the *National Spatial Planning Act* in 1950, which possessed the characteristics of the basic law or organic law for land development. Japan's five National Comprehensive Development Plans were established in accordance with the law. In addition, more than 60 laws and regulations have been formulated for the development of large urban areas, industrial zones, underdeveloped areas, mountain village revitalization and so on, such as the *Development Act of Capital Circle* and the *Industrial Distribution Control Act of the Capital Circle*, and to implement the National Comprehensive Development Plan[11]. The French *Land Resources Planning Act* is the basic law of construction. The *Planning Reform Law* 1982 proposed "planning contracts", which take two forms, one is the contract between the state and the region with the purpose of implementing priority projects for economic development and land improvement, and the parties to the contract pledge to provide the funds necessary for the implementation of specific projects. The second form is the contract between the state and the state-owned enterprise, which does not stipulate any quantitative indicators, but mainly determines the medium- and long-term development plan and priority construction projects of the enterprise, finalizes the general strategic policy jointly formulated by the two parties, expounds the objective in terms of social benefit that the state expects the enterprise to achieve, and clarifies the obligations of the state to the enterprise in terms of financial subsidies, investment and external environment. The planning contract has legal effect on both parties upon signature. In 1995, the French Parliament passed the *Act Guiding National Land Improvement and Development*, including the formulation of a national outline of land improvement, the creation of a national committee for land improvement and development, the setup of new pilot administrative district(s), and the establishment of new industry development fund, which boosted the state financial subsidies and policies primarily favoring key areas.

The United States does not have straightforward regional planning laws, but the *Regional Rehabilitation Act* and the *Urban Growth and Community Development Act* undertake certain regional management powers. When it comes to the overall interests of the country and the cross-state development tasks, the federal government usually

[11] Zhang Jifeng. *Theory of Comprehensive Land Development of Japan*. Beijing:World Knowledge Press, 2004.26.

formulates relevant acts to stipulate the tasks, objectives, and economic policies of the planning, demanding, or encouraging the departments and states involved to take active part in the organization and implementation of the planning. Generally, a substantive body is set up to take the responsibility of organizing, coordinating, and implementing interstate planning, and matters belonging to the states are legislated in accordance with federal legislation, the general requirements, and funding situation, and a special agency will be established for the implementation.

3.3.4.2 Economic Means

Economic means are most important for developed countries to guide space development and protect space resources. The implementation of the plans is mainly ensured through the government's public investment, economic subsidies, funds, tax policies, etc.

The "structural funds" of the EU has long been proved an effective tool for facilitating program implementation within EU. Since the 1950s, structural funds have, despite the many changes in the organization of the European Union, played an irreplaceable role in guiding industrial layout, revitalizing declining areas, and promoting coordinated regional development.

In 1965, Japan stipulated that in the basic construction plan of new industrial cities or areas for special industrial preparation, public facilities such as housing, roads, harbors, health, and education could get certain subsidies from the national treasury and preferential interest on loans. In 1972, it was stipulated that all enterprises established in the designated inducing areas could be exempted from taxes according to law.

In Germany, the *Basic Law* stipulates that the federal government is responsible for building and managing federal infrastructure, social security, cross-regional development, and national economic development and adjustment. States with high per capita incomes are mandated to help states with low per capita incomes through grants, and the federal government subsidizes financially insufficient states through vertical grants. The federal government has put aside 20% of its budget for investment in social infrastructure. Over the years, German land authorities at all levels have at their disposal certain amount of subsidy funds earmarked for enterprises to guide their construction in accordance with the requirements of the land resources planning.

France has also set up various special funds, such as the land improvement fund and the rural improvement and development fund, to facilitate the implementation of land improvement projects. Besides, certain subsidy measures have also been adopted, such as the offering of regional employment subsidies of 3500–5000 francs for each job created in priority development areas where investment is made.

3.3.4.3 Administrative Means

(1) Governing Body with Clear Responsibilities

Japan has a strong top-down governing body for land resources planning. The Ministry of Land, Infrastructure, Transport and Tourism is the main body responsible for implementing land resources planning, regional planning, and urban planning, and each functional department is responsible for the preparation of special plans. For instance, the Ministry of International Trade and Industry is responsible for planning of industrial development and re-allocation, the Ministry of Transport is responsible for transportation planning of roads, railways, ports, and terminals, and the Economic Planning Agency oversees the planning of national economic policy. The Ministry of Land, Infrastructure, Transport and Tourism is a powerful planning mechanism that integrates the preparation, management, and implementation of land resources planning, which played a key role in the rise of Japan's postwar economy and became a fine model for regional spatial planning and management in the world.

France has long adopted the highly centralized administrative system. In the early 1960s, its government re-organized its 96 provinces into 22 regions to improve the efficiency of land management, and in 1982 the regions became national-level administrative regions. The Department of Land Improvement and Urban Planning under the Ministry of Public Works, Housing, Transport and Tourism is responsible for national urban planning and the formulation and management of national macro-level planning. The coordination between the central government and local departments is mainly through the coordinating committee of urban development ministries and commissions and the Urban Development Fund. Branches are set up in the regions and provinces to guide and coordinate the planning of cities and towns, to solicit the opinions of local authorities and the public, and to supervise key national construction plans related to municipal infrastructure, public and disaster prevention projects.

The U.S. federal government does not have a dedicated regional planning body. State-level planning and management is mainly the charge of the Urban Affairs Commission set up within the Office of the President. The Ministry of Housing and Urban Development of the federal government provides financial and technical support to regional and urban organizations and local municipalities and counties for their work related to various planning. The federal government exercises a certain number of powers primarily through the allocation of federal funds. It is worth pointing out that although there is no institution to implement regional planning and exercise related functions in the US, there are many specialized organizations dedicated to solving various regional problems or "governments of special zones" that exercise certain single function. For example, South Coast Air Quality Management District (AQMD) and the Southern California Association of Governments (SCAG) of the Greater Los Angeles Area, whose main function is to coordinate regional issues, specifically through the allocation of annual funding worth billions of dollars to implement the plans.

The UK Department for Environment is the highest regional and urban planning body in the country (with the Office of Urban and Rural Planning and the Planning

Supervision Commission set up within it) and has planning departments at the county and city levels responsible for organizing and guiding regional spatial planning at all levels. With the rapid progress of privatization since the 1980s, the interests of private enterprises and many interest groups became more and more dependent on the development of the region where they were located, as a result, they constantly demanded participation in regional planning. In England, a new type of regional governing body emerged in the form of regional association. There is also the network of government regional offices managed by the central government and bodies managing EU regional projects. Regional management has stopped being the action of the government alone, and became a process extensively participated by the private sectors and related stakeholders.

(2) Management through Project Approval

Generally, the development of certain regions, certain projects and enterprises is encouraged or restricted through approval, issuance of licenses, signing of contracts, etc. Germany restricts and discourages the construction and development of certain enterprises mainly by exercising the power of land management and review. In Japan, every ministry (agency) is invested with certain approval powers and able to use incentive policies in terms of loans, taxes, and subsidies. In France, only the projects that are deemed unsuitable to implement elsewhere and only after they are reviewed and granted license by the "Relocation Commission", can they start construction in Paris.

3.3.4.4 Social Means

The public are engaged to gain broad social support and ensure and supervise the implementation of regional plans. Take the preparation of the national spatial order planning of Germany for example, its draft was printed out and distributed to governments at all levels and relevant departments for comments and was made available to the public for their opinions; after being revised and passed by the parliament, it was officially issued by the government and disseminated to the public. Enterprises and relevant entities involved should be familiar with the requirements and principles of land resources planning and observe them conscientiously in their work. With the land resources planning of Japan, opinions of various departments and regions were solicited several times during its formulation, and it was submitted to the Cabinet Council for approval after a consensus was reached. After its publication, vigorous efforts were made to raise public awareness of it, and the government, local public organizations, organizations, and residents of civil society were all called upon to participate in the development and construction of the national land.

3.3.4.5 Other Means

(1) Periodic Inspection System

Every two years, the German government makes a report on the basics of national land resources planning to the House of Parliaments. During the implementation of its 3rd Comprehensive Plan, Japan decided to halt and revise it given the significant change in international situation (the oil crisis) which made the original plan difficult to carry on. The Dutch central government established a supervisor system of land resources planning, which put one supervisor in charge of supervising the implementation of land resources planning of two or three provinces, and they have the right to make planning recommendations to provincial land resources planning commissions and municipal government departments to ensure that provincial and municipal planning is in line with the national policy.

(2) Adopt Advanced Technical Means to Lay the Scientific Foundation of Planning

Databases and computers are widely employed in the regional planning in foreign countries, and data analyses including regional development and structural changes are adopted. In Germany, the regional planning bureaus of all states have established database and computer intelligence collection and processing systems for regional planning, making it possible to retrieve data at any time for quantitative analysis of the current situation and future developents. This way, the qualitative analytical conclusions are based on many quantitative analyses.

(3) Dedicated Planning Advisory Bodies in Place

In Germany, regular administrative bodies of land resources planning are set up at the levels of federal, state, regions, and cities counties, which are responsible for preparing the revisions to the plans every five years. There are also independent research institutes of land resources planning, such as the German Academy of Land Research and Land Resources Planning, whose mission is to identify and propose solutions to problems in spatial and environmental development at an early stage. The Land Development Research Institute in Korea specializes in the research and formulation of land resources planning.

3.3.5 Outcomes of Spatial Planning

The land resources planning or regional planning under the planned economy system had played an important role in the first 50 years of the twentieth century. Since the 1960s, however, the following problems emerged gradually: excessive and rigid management and control, lack of vitality of local areas and enterprises, and high input and low output. Added to these the serious imbalance between agriculture, light industry and heavy industry, and the lack of drive for upgrading the industrial structure, economic development since the 1980s showed a marked decline as a result, and eventually led to the collapse of the planned economic system. It is worth pointing out that the national industrial spatial distribution and the organization of spaces using "regions" as the unit under the planned economy have had played a positive role in improving the country's industrialization and the people's living standards in a short period of time.

National spatial planning or regional planning in a market economy plays a different role due to the diversity of its categories. The series of land resources plans developed by Japan after World War II played an important role in the country's postwar economic recovery and rapid improvement of overall competitiveness, and it is still an important part of its national policy today. South Korea's national spatial planning since the 1960s has contributed to solving the host of problems in the period of rapid industrialization, such as uneven regional development and uncoordinated development between urban and rural areas. The five national spatial planning of the Netherlands since 1960 contributed majorly to enhancing the country's status in Europe and improving the quality of living environment. The regional planning of New York in the 1920s and the Tennessee Valley regional plan in the 30 s have had an important impact on the development of regional space in the United States as it laid the foundation for the state-based model of regional planning for the implementation of the federal government's national construction plan. In England, planning at the regional level has been continuous since the Greater London planning after World War II, ranging from the construction of new cities to the revitalization of declining areas, regional planning of the southeast area in the new century, and the national-level regional planning under the guidance of relevant policies—all of these indicate that macro-level planning policies and guidelines have always been playing an important role. And the recent discussion on the adjustment of zoning has prompted people to gain a new understanding of the role of regional planning.

To sum up, spatial planning is an important means for the state to coordinate economic, social, environment and regional development in a comprehensive manner under the market economy, and a process through which public policies are formulated. It has an important role to play in guiding industrial development and managing space resources. Judging from the cases available, the major technical methods include: (1) With global development in mind, cultivate cities and regions that are capable of participating in international competition, and improve national competitiveness; (2) Build a balanced urban spatial structure guided by industrial development; (3) Attach importance to the coordinated development of regions, to be specific, solving the problems of densely populated areas packed with economic activities, and at the same time promoting the development of relatively backward areas; (4) Make active efforts in guiding public investment, strengthening infrastructure construction, and providing service systems that is equally accessible to all; (5) Protect and utilize such space resources as the land, water, energy, oceans, minerals, and forests in a sensible way to achieve sustainable development; (6) Facilitate the implementation of planning through such means as legislation, finance, public engagement and so on, and guarantee the financing for space development and guidance.

With spatial planning at the national level, efforts should be made to handle well the following relations: (1) The relation between the government and the market. In a market economy, planning is an important means for the government to carry out macro-control and intervention in the market, which is by no means to disrupt the laws of the market, but to make up for what is lack of the market. (2) The relation between the central and local governments. Taking the overall development of the country into consideration, the plan needs to coordinate the complex relationship between the central government and various parties and maximize the interests of the country. (3) Local-to-local relations. The planning plays the role of coordinating various local interests in the development and utilization of space resources, thus ensuring regional economic development and sensible use of resources. (4) The relation between industries and urban space. Handle well the relationship between industrial development, population migration and urban spatial expansion. (5) The relation between development and protection. While promoting the development and utilization of resources, we should bear in mind the protection of the ecological environment and the prevention of crises. (6) The relation between urban and rural areas. Through planning, we should handle well the relations between cities and villages, between developed and backward areas, and between densely and scarcely populated areas.

3.4 Understanding of Government Intervention According to the Theories of Political Economy

The evolution of regional spatial structure in the era of globalization is strongly affected by economic forces, but the development of cities and regions is not a pure economic process. From the perspective of national competitiveness, the country still has considerable political significance. Therefore, an analysis of the theory of government intervention from the angles of economics and political science has important practical significance for the practice of socialist market economy that we are currently engaged in.

3.4.1 Guidance of Non-equilibrium Theory to the Order of Spatial Development: Relations Between Pioneers and Latecomers

3.4.1.1 Balanced Development Theory and Its Limitations

From the outset of their study, development economists have been concerned with the role of space in development theory. The earliest classical theories were equilibrium development theories based on Harrod-Domar's neoclassical economic growth model, including Rosenstein-Rodan's theory of "Big Push" and Nurkse's theory of balanced development. The "Big Push" theory "reveals the underlying conditions of pecuniary externalities for development" [20], while the theory of balanced development argues that backward countries are prone to fall into the vicious circles of insufficient supply (low productivity—low income—low savings—insufficient capital supply—low productivity) and insufficient demand (low productivity—low income—insufficient consumer demand—insufficient investment demand—low productivity). The key to breaking the vicious circles is to foster development in a balanced manner in all industries and regions at the same time, promote coordinated development of various industries and sectors, and improve the level of supply and demand. Therefore, the theory of balanced development attaches importance to the connectivity and complement between industries and regions and advocates balanced allocation of factors of production among industries and regions. This idea was practiced in the planned economy era of the former Soviet Union and China.

The balanced development theory inherits the main ideas of neoclassical economics and aims to promote coordinated industrial development and narrow the gap of regional development. But in nature, it pursues fairness of economic development at the expense of efficiency, added to this the difficulty to apply it, the theory has been widely criticized. Since a common problem faced by developing countries is the scarcity of capital, those countries do not have the conditions to pursue balanced development [21].

3.4.1.2 Non-equilibrium Theory and Spatial Development Policy

Based on the reality of scarce development resources in developing countries, the non-equilibrium theory stressed the importance of prioritizing the development of key regions and key sectors to drive the overall economic development. This theory is useful in guiding the order of space development.

(1) Theory of Growth Pole

It was first proposed by the French economist Perroux, and later enriched and developed by the French economist J.B. Boudeville, the American urban economist John Frishman, the Swedish economist Gunnar Myrdal, and the American economist A.O. Hischman to varying degrees. According to the theory, "for the country, balanced development is only an ideal that is impossible to achieve in reality. Economic growth usually started from one or more 'growth centers', and then gradually transmitted to other sectors or regions. Therefore, a specific geographic space should be selected as the 'growth pole' to promote the polarization of spatial economy. The purpose of prioritizing the development of certain areas is to facilitate the birth of agglomerative economies effect through concentrated use of the factors of production. Agglomeration and concentration bring about the saving of production factors and make the allocation of resources more reasonable."[12] Rodwin first applied the growth pole theory to regional planning in the 1960s, and it has since then become the most widely used development model in regional planning in developing and underdeveloped regions.

---

[12] *Exploration of the Growth Pole Theory and Regional Economic Development Strategy in Underdeveloped Areas*, Zhang Jinpeng. *Modern Economic Science*. 1999 (NO.6).

(2) Point-Axis Development Theory

Proposed by the Chinese geographer Lu Dadao as the extension of the growth pole theory, the point-axis theory gives equal emphasis to the role of the "points" and the role of the "axes" between "point" and "point". "Point" in here refers to the central areas at all levels, which play the role of driving regional development, and "axis" refers to the relatively densely populated industrial belt formed by connecting several central towns of different levels in a certain direction[13]. In 1984, Lu stated that "Whether it is a country or a region, in its early stage of large-scale economic development or the period immediately after that, development should be concentrated in a few points or areas due to the limited funds, material resources and regional infrastructure. This way, a higher rate of economic growth can be achieved, and it would be impossible to achieve if investment were to spread in all places with no agglomeration effect generated."[14] The point-axis theory holds that industry is always first concentrated in a handful of cities or enterprises with better conditions, in a pattern of points. Then, to fulfill the needs of exchanging factors of production, points are connected to form axes through such lines as transportation, power supply, communication, water supply, etc. It is thus necessary to give priority to the development of the points and axes with better conditions to drive the overall development of the region. With the development of the economy, the axes would in time extend to the less developed areas, thus forming new growth centers, and the growth centers and development axes of different levels would then form a socioeconomic spatial network[15]. As a result, balanced and coordinated development of the regional economy will be eventually realized through the development of such networks.

(3) Theory of Circular and Cumulative Causation

The Swedish economist G. Myrdal proposed the structural theory of "geographical dual economy" by employing dynamic non-equilibrium analysis and structuralist analysis. In essence, the theory of circular and cumulative causation means that developed regions would produce "echo effect"[16] and "diffusion effect" on backward regions. The echo effect tends to realize the "Matthew effect", that is to make developed regions more and more developed, and backward areas more and more backward; the diffusion effect is conducive to driving the development of backward regions. [22] observed that the echo effect is much greater than the diffusion effect under the influence of market forces. "The market forces tend to increase rather than decrease regional differences" [22], thus creating a cumulative circular trend that is the source of regional economic imbalance. According to the theories above, Myrdal and others believe that government intervention is a necessary means to promote the

---

[13] *Regional Development and Its Spatial Structure*, Lu Dadao. Beijing: Science Press, 1995. 137.

[14] "The Scientific Basis of China's General Industrial Productivity Layout, 2000", Lu Dadao. *Scientia Geographica Sinica*, 1986 (Vol.6, No.2).

[15] "Analysis of the Formation Mechanism of Point-Axis Spatial Structure System", Lu Dadao. *Scientia Geographica Sinica*, 2002 (Vol.22, No.1).

[16] The echo effect refers to the return and accumulation of factors of production to the development pole.

coordinated development of regional economy. When certain regions have accumulated development advantages, the government should adopt the strategy of uneven development and give priority to the regions with stronger growth potential, and then drive the development of other regions through diffusion effects. He also cautioned that the difference in development from region to region should be kept within certain limits, and the government needs to take measures to prevent the gap between the rich and the poor caused by the cumulative and causal cycle from widening indefinitely.

(4) Theory of Unbalanced Growth

Sharing some similarities with Myrdall's theory, A. O. Hirschman's theory of unbalanced growth pointed out that development in the North (the regions of growth) will produce the dual effects of "trickling down effects" and "polarizes effects" on the South (the backward regions). The "trickling down effects" leads to increased purchasing power of goods and investment of the North to the South, as well as hidden unemployment and emigration from backward areas. The "polarizes effects" would draw key factors of production to the North, thus inhibiting economic activity in the South and widening the economic gap between the North and the South. Having a deeper understanding of imbalance of regional economy than Myrdall, Hirschman advocates that unbalanced strategies be used as the ideal means to develop economy. He stated clearly that development is an evolutionary chain process unbalanced in nature; strong forces of economic growth are bound to form around the points where they were first produced; the emergence of growth points or growth poles proves that inequality of growth in regions is the inevitable result and prerequisite of growth itself; and the cumulative concentration of economic growth is the inevitable result of development. Therefore, he advocates that the purpose of development is to make sure that imbalances exist, rather than eliminating them—"the infrastructure of backward areas being slightly insufficient for their production activities help those areas to make sensible use of their scarce financial resources", and the task of development policy is to maintain regional imbalances and the gap in potential.

(5) Gradient Development Theory

The gradient development theory was originated from the "industrial production life cycle theory" developed by Harvard University professor Raymond Vernon and others and later summarized and developed by Hoover and Chinese economists into the theory of regional economic development. According to the theory, "the development of various regions of a country often has a sequence of development and a gradient in development level. The level of gradient is mainly determined by the advantages and disadvantages of the industrial structure of the region, especially the stage of the leading industrial sector in the industrial life cycle. The region high on the gradient is the birthplace of new products, new technologies, new ideas and new ways of organizing and managing production. With the passage of time and the decline of the life cycle, the renewal of industrial structure and the development of regional economy were gradually phased out from high-gradient areas to low-gradient areas. The graduation process on the gradient is the result of the combined effect of polarization effect and diffusion effect generated dynamically. In the order of regional economic development, we should first support and promote the development of the economy of high-gradient areas, to achieve better economic returns to drive and facilitate the development of low-gradient areas.“[17]

(6) Urbanization Theory of New Classical Economics

New classical economics was founded by Yang Xiaokai in the 1990s. Using inframarginal analysis, he synthesized classical economics more deeply and thoroughly than did the neoclassical economics. In the study of urban problems, it abandoned the concept of economies of scale and used the concept of specialized economy as the tool of research. According to new classical economics, centralized transactions, with its merit of saving transaction costs, gave birth to cities. Large, medium, and small cities would form of their own accord in the market, and this stratification is the result of the trade-off between the efficiency and cost of centralized transactions. The pattern of division of labor in large, medium, and small cities or the status and role of each city depend on the size of its division of labor network or the level of division of labor. With the division of labor become more specialized, the market will gradually form a sound urban system, which is the result of a compromise between the efficiency (advantages) and costs (disadvantages) brought by centralized transactions, and the distribution of large, medium, and small cities of different scales is inevitable. New classical economics upholds that in spatial development, neither large cities nor small cities should be the sole focus of attention; only the coordinated development of large, medium, and small cities is the market-oriented choice[18].

3.4.2 Prevalence of Moral Hazard Requires the Intervention of Central Government: Relationship Between the Central and Local Governments

3.4.2.1 Theoretical Basis for the Prevalence of Moral Hazard in Local Governments

Institutional economics holds that in modern society, economic man with “perfect knowledge” do not exist, and the interaction between people is subject to two deficiencies—“vertical uncertainty” and “horizontal uncertainty”[19] [23], and imperfect information is the norm, and there are always differences in the amount and quality of

[17] *Development Strategy of China’s Regional Economy*, Zhu Houlun. Social Sciences Academic Press. 2004.

[18] See *Specialization and Economic Organization: A New Classical Microeconomic Framework*, Yang Xiaokai and Huang Youguang. Economic Science Press. 1999.

[19] “Vertical uncertainty” refers to uncertainty about the future caused by a lack of understanding of the future, and people’s actions must be based on speculation about the future. “Horizontal uncertainty” refers to uncertainty about the precise characteristics of resources, potential trading partners, etc., especially when people need to enter cooperation, and it is often difficult to judge the possible behaviors of those agents.

information owned by different participants in economic games, rendering specialization necessary. The problem is that with economic behaviors, due to the asymmetry of information between the principal and the agent, it is difficult for the former to regulate the behavior of the latter. Under the bureaucratic system, the more tiers of the entrusted agent, the longer the chain, and the greater the possibility of information decay and moral hazard, hence the increased difficulty for the principal to discover and correct in time the opportunistic behavior at the micro level.

Extending the use of the "economic man" assumption in the economic field to the political process, Buchanan analyzed the self-interested tendency of the government, providing an important basic premise for the analysis of the relationship between the central government and local governments. The central and local governments are the same in nature and both enforcers of state power, but due to their being at different levels, each shoulders different responsibilities to the state. As the government that represents the country, the central government shoulders the responsibility of safeguarding the stability of the regime, national unity, and sustainable economic and social development of the country by giving priority to the overall interests of the country. Although having something in common with the central government, local governments have interests of their own and would pursue their own interests at the expense of the overall interests when the former are large enough for them to break through the constraints of government norms, rendering the entire social and economic life caught in the serious "tragedy of the commons" under new conditions; the overall interests would be eroded by local interests and organizations [24]. In fact, problems such as the irresponsibility of local governments to higher authorities and the public, and the difficulty in coordination and cooperation are common problems worldwide[20].

3.4.2.2 Prevention of Moral Hazard in Local Governments

Considering the moral hazard issue in the principal-agent relationship, economists of the new institutional economics believe that the principal should exercise control over the agent at various stages including pre-, during, and post- the process. Pre-control refers to the determination of the norms and rules regarding agent's behaviors and the boundaries of power in advance through forward planning and preventive measures, thus restraining the agent's opportunistic behavior as much as possible. Control during the process refers to forcing the agent to enhance the transparency of its execution, and to inspecting and monitoring the agent's behavior from the perspective of procedures. Post-control refers to evaluating the agent's performance, inducing the agent to align its interests with that of the principal through reward or punishment. Of the above, pre-control is the key link of control.

From the perspective of the methods of control of the market economy governments, the administrative control method of western developed countries has undergone major changes. After the rise of Keynesianism, the idea of macro-control was

[20] See *Distributed Public Governance* (Chinese translation), OECD. CITIC Press, 2004. 22–26.

generally accepted by western governments who adopted macro-management, and administrative guidance and control to regulate economic behavior. But to overcome the rigidity and passivity of conventional administrative control, the governments of western developed countries opted for the method of administrative guidance. Through guidance and encouragement, this method not only gives play to the initiative of the central government in administration, but also generates the enthusiasm of local authorities and improves the efficiency of government actions.

3.4.3 Externality Requires Planning to Function in a Regional Context: Local-to-Local Relations

3.4.3.1 Conflicts and Games Between Local Governments

In addition to the conflicts with the central government, the "economic man" nature of local governments is also manifested in the relations among themselves. To be specific, local governments are characterized by "their ability to influence or mobilize social resources by way of state power within the scope of law. In this way, they would achieve good performance, bringing benefits (such as income and income compensation) to members of political interests [25]. It is on this basis that various forms of competition and conflict between local governments formed.

Competition between local governments is exclusive in nature; each seeks to maintain its "turf" by instinct. From the perspective of expansion, once its physical space is determined, local governments have only one way to expand outward—the infiltration of economic power to obtain more share of the cake. History, be it of the world or the countries or regional development, is always laden with both cooperation and conflict [26].

3.4.3.2 Externality and "Prisoner's Dilemma"

Regional interest groups have limited rationality when they are pursuing the maximization of their own interests in competition. Their behavior has externality just as the natural person and the enterprise, which means that the activities of any interest group always have a positive or negative impact on other interest groups. The existence of inter-regional externality is the root cause of the contradiction between regional economic interests and between regional interests and overall interests.

The "prisoner's dilemma" of public goods supply: public goods are indivisible and natural monopoly in production, and non-rejective and non-competitive products in consumption [27]. The nature makes it difficult to prevent "free rider" in the field of public goods supply. For local governments, enjoying public goods but not paying the cost is the best option, the second-best option is to enjoy the public good while paying the cost of their share, the third best is not to enjoy the public good and pay no cost, and the worst scenario is to pay for the free rides. The free rider dilemma is one of the most common "prisoner's dilemmas" in the provision of regional public goods. In this dilemma, when public goods are insufficient, local governments would expect financial allocation from higher levels of government or central government, and then compete for bigger share of the allocation, while the weaker areas simply adopt an opportunistic attitude, waiting for the strong regions to provide regional public goods and free ride.

The "tragedy of the commons" in the use of public resources: it is a common phenomenon disclosed by British scientist Hardin. According to him, public resources are often over-exploited in the absence of rules. He postulates an open pasture where overgrazing led to gradual depletion of the field when the ranchers raised more cattle than the maximum capacity the field could hold. Hardin believes that without a law limiting the number of cattle herded by the ranchers, the tragedy cannot be avoided. Because although the destruction of the field will eventually harm everyone's interest, each individual rancher only cares about the increased benefit of adding one more cattle, so he will increase the number of cattle as much as possible. Eventually, each rancher's very action of maximizing his own profit ended up harming the interests of everyone, including himself. In aggressive competition, local governments try to draw more investment by improving the soft and hard environment such as improving infrastructure and providing preferential policies. However, due to the lack of rules, local governments vied with each other in carrying out large-scale infrastructure construction and preferential policies at all costs, which ended up harming the interests of all competitors.

3.4.3.3 Solution to Prisoner's Dilemma and Intervention of Central Government

Either the insufficient supply of regional public goods or the tragedy of the commons in public resources utilization will harm the interests of the region. The solution to the prisoner's dilemma depends on the establishment of cooperative systems and mechanisms. Game theory identifies the conditions for the existence of cooperative games from the game process itself, that is, the same gamers keep repeating the game, and cooperative solutions will appear. According to the views of new institutional economics, institutional change is a response to institutional non-equilibrium, and it is a change in the system from non-equilibrium to new equilibrium. To solve the problem of insolvable conflict due to high transaction cost or even the inability of transaction at all, parties to the conflict often resort to a third party for it to coerce or influence one of the parties to settle the conflict through equal consultation. The introduction of a third party in this way significantly reduces transaction costs, so that conflicts that could not have been resolved through negotiation can be now, turning non-cooperative games into cooperative ones.

From the perspective of the common interests of the region, there is indeed the possibility of cooperation between local governments, the representatives of local interests, through consultation, negotiation, and exchange of information, but the regions are not all that amicable in their dependent and complementary relations, and there is no equality among the parties in the game. The relatively backward region often appears to be more passive in negotiations due to the strong constraints on them, while the stronger party is obviously the one taking the initiative as its ability to choose can turn into an "effective" voice in negotiations. When such internal contracts are difficult to reach and regional interests cannot be coordinated internally, it is imperative to introduce a third party, that is, the central government, to make up for the lack of effective supply of the system through national laws, regulations, and plans. Rather than favoring one party over the other, government intervention aims to foster a win–win arrangement of two or more parties, i.e., the formation of an equitable mechanism of the region. Government intervention is mainly carried out through the following means: clear definition of property rights between regions from the outset (such as water rights, pollution discharge rights, etc.), the direct supply of regional public goods of national significance (such as regional infrastructure), the provision of regional cooperation systems, and the coordination of conflicts and disputes between regions. It is safe to say that spatial planning is the main aspect of it.

3.4.4 *Equal Starting Point and Narrowing Regional Gaps Through Planning: Relations Between Developed and Backward Regions*

3.4.4.1 Economic Development and the Disadvantages of Developing Countries as the Latecomers

Economic development is a concept of two aspects: quantity and quality. Although growth plays an important role in development, the two are not necessarily linked. The prescriptive nature of the concept of development is embodied in changes in the economic structure, including changes in the structures of inputs and outputs, changes in the living standards of the people, and improvement in distribution. If, during economic growth, there is no change in the mode of production, the industrial structure is not optimized, the living standards are not improved, and the inequality in income distribution is even worsened, then the growth will not bring real economic development [28].

Under the current environment of fierce competition in economic globalization, the governments of developing countries are under great pressure for development because economic development has become the main cornerstone upon which they establish political legitimacy. However, the common problems faced by these countries are their failure to satisfy the basic needs of all members of society, and equally distribute the fruits of growth and promote social progress. The root cause for this is that "the too hasty nation-building and the overly urgent pressure of modernization deprived the developing countries of the time necessary for them to build the institutions which would have helped them systematically treat the ills arising from the contemporaneous political building and modernization one by one" [29]. When latecomers to economic development try to catch up with their developed peers, they would first imitate the industrialization model, then their economic systems, and last the legal systems and behavioral norms [30].

Unlike developing countries, early developed countries had the time to improve and perfect their institutions thanks to the slower development over the course of centuries of modernization. For developing countries, however, "the accumulation and worsening of economic and political inequality in a short period of time will lead to serious conflicts of interests, social conflicts and political turmoil" [29]. So, the imbalanced development constitutes a serious challenge to the political stability of developing countries, and to solve this problem is as important as economic growth.

3.4.4.2 Capability Poverty and Equality of Opportunities

It should be noted that economic disparities are only the symptoms of development inequality, and deeper hidden is the inequality of political power and the lack of development capacity. "Relative deprivation of income produces absolute deprivation of capability" (Amartya [31]). The consequence of capability poverty is the solidification or even exacerbation of inequality. From the perspective of economic-political relationship in regional development, "when economic disparities lead to political problems, unequal distribution of political power will also lead to unequal economic development... The political impact of regional development is often overlooked when explaining regional inequalities. However, it is obvious that power relations in terms of space have a direct impact on regional development. And regions that have a say in central decision-making are better able to secure more resources for their own development than regions that have no weight in it."[21] The lack of development capacity has led to the relative solidification of social stratification and the poor groups' continuous replication of themselves. Impoverished areas are caught in a vicious cycle of "poverty-lack of capacity-poverty". The key is the extent to which our policies can eliminate capacity poverty and give everyone equal opportunities for development. As Rawls argues, the ethical choices that a society should have must consider the social opportunities of the most unfortunate people in society, and only a society of equal opportunities is a fair society [32].

3.4.4.3 Equal Starting Point and Responsibilities of Central Government

Equal opportunity means, first and foremost, equality of starting points, that is, the balance and rationality of the starting point of economic competition. In the unified relationship between equality of starting point, process and result, equal starting

[21] See The Role of Polities in Regional Inequality: The Israeli Case, Yuhuda Grandus 388–403.

point plays a decisive role and is the fundamental guarantee of the equality of the other two. If there is no equality in starting point, or worse, there is polarization, then the competition process cannot be equal, and an unequal competition will amplify the inequality effect of the starting point, which eventually lead to greater injustice and polarization in the result. "If we ignore the state of the starting point, and simply emphasize the so-called fair competition of the process, or even try to eliminate social polarization merely through the social welfare system and the government's redistribution mechanism, the result will only be a failure or even be counterproductive." This is not only a matter of theoretical logic, but also a true summary of the policy practice of developed countries" [33]. From this point of view, we cannot rely entirely on the improvement of the social welfare system and income redistribution to narrow regional disparities, the key is to strive to improve the development conditions of backward areas so to achieve equality in starting point. Just as Huntington proposed in solving the political disorder caused by the urban–rural gap that rebuilding political stability requires the formation of an alliance between urban groups and rural people, the stability of national politics also requires the formation of some kind of alliance between different regions, the interests of backward regions should be incorporated into national politics, and the masses in backward regions should also be encouraged to participate in politics. Therefore, governments should play a leading role and assume more responsibilities in matters related to social equity, narrowing the gap between the rich and the poor, and supporting vulnerable groups, including human survival, health, education, etc. And in here also lies the major role of national-level spatial planning.

3.5 Summary

The development of urban space involves many aspects. From the perspective of the construction of human settlement environment, urban development requires comprehensive thinking and vision at the macro level. From the perspective of economic globalization, it should take global competitive environment into consideration. From the perspective of planning, it is a means of the government to guide and regulate national and regional development. So, instead of a mere technical process, it is also a political process. From the theoretical perspective of economics and political science, it is imperative for the state to intervene in spatial development, which can prevent the infinite amplification of market forces, echo effects and so on, especially so for countries with low starting points and faster development. From the perspective of national competitiveness, the country still has considerable political significance in the era of globalization. The development of urban space should fully consider the ecological environment, economic foundation, social and cultural conditions, and the development of cities and regions should be compatible with the construction of human settlements.

References

1. Wu L (1982) Proceedings of Urban Planning
2. Hall P (2003) Cities in civilization: culture, innovation and urban order[J]. J Irish Urban Stud 2:1–14
3. Amin A, Thrift N (eds) Globalization, institutions, and regional development in Europe, Oxford University Press (1994)
4. Wang J (2001) Comparison and reference of land planning systems in japan and ROK. China Land Sci 3:45–48
5. Morgen S (2001) The agency of welfare workers: negotiating devolution, privatization, and the meaning of selfsufficiency[J]. Am Anthropol 103(3):747–761
6. Nolan P (2001) China and the global business revolution. PALGRAVE publication. 29–41.
7. UNCTAD Trade and Development Report (1999) United Nations conference on trade and development
8. Alden DL, Steenkamp JBEM, Batra R (1999) Brand positioning through advertising in Asia, North America, and Europe: the role of global consumer culture[J]. J mark 63(1):75–87
9. Wang ZL (2003) Investment report of multinational corporations in China. China Economic Press, Beijing, pp 15–23
10. Krugman P (1991) Increasing returns and economic geography. J Pol Econ 99:183–199
11. Porter ME (1998) The Adam Smith address: location, clusters, and the 'new' microeconomic of competition. The National Association of Business Economist
12. Hall P (2002) Urban and Regional Planning, Fourth Edition, London: Routledge
13. Sassen, S. (1991) The Global City: New York, London, Tokyo. Princeton University Press, Princeton.
14. Gottmann, J (1961) Megalopolis: The Urbanized Northeastern Seaboard of the United States. Twentieth Century Fund, New York
15. McGee TG (1991) Presidential address: Eurocentrism in geography-The case of Asian urbanization. Canadian Geographer/Le Géographe canadien 35(4):332–344
16. The mega-urban regions of Southeast Asia. UBC Press, 1995.
17. Montagnon P (1997) Total of Asia's megacities set to double. Financial Times, April, 21, 1997
18. Yang WM (2003) Theoretical exploration of planning system reform. China Market Press, Beijing, p 6
19. Zhang JF (2004) On the comprehensive development of land resources in Japan. World Affairs Press, Beijing, p 26
20. Krugman P (2000) Development, geography and economic theory. China Renmin University Press 14.
21. Han FQ (2004) Regional disparity: government intervention and public policy analysis. China Financial and Economic Publishing House, Beijing, p 30
22. Myrdal G (1957) Economic theory and underdeveloped regions (English Edition). Duckworth Books, London, p 26
23. Ke W, Shi M (2008) Institutional economics: social order and public policy. Beijing: The Commercial Press
24. Huang YY, Chen CC (2001) Behavioral analysis of local government institutional innovation. Local Government Administration (7)
25. Gao JM (2005) Protective competition and collusion mechanisms of local governments. www.cenet.org.cn
26. Zhang KY (2005) Regional economic policy. The commercial press 81
27. Yu YD et al (1997) Western Economics. Economic Science Press, Beijing, pp 222–223
28. Zhang PG (2001) Development economics. Economic science press pp 38–39.
29. Hu WJ (2005) The political logic of equilibrium. Chongqing Publishing House, Chongqing
30. Hu YT, Yang XK, Sachs J (2003) Economic reforms and constitutional transition. Q J Econ (4)
31. Sen A (2002) Development as freedom (Chinese Edition). Renmin University of China Press, Beijing, p 87

32. Rawls J (2000) Lectures on the history of moral philosophy. Cambridge: Harvard University Press
33. Duan XM (2004) Equal opportunity, impartial competition and common well-off. J Sichuan Polytech Inst (Social Science Edition)

Chapter 4
Review of Urban Spatial Development History in Modern China

4.1 First Practices of Regional Planning in Modern China

4.1.1 "Local Autonomy"-Styled Regional Planning Pioneered by Zhang Jian

The Opium War in 1840 ushered in the era of semi-feudal and semi-colonial society in China. The capitalist invaders forced the Qing government to sign a series of unequal treaties, helping them open the door to China, which led to the development of commerce and trade in the coastal port cities. Against the background of "Westernization Movement", national capitalism also developed to a certain extent, leading to the rapid development of the industrial and commercial cities along the coast. The road of urban and regional planning in this era was difficult and tortuous. Although colonial cities have some sort of urban plans, most of them bore the stamp of suzerain. As far as regional-level planning is concerned, the "local autonomy"-styled regional development carried out by Zhang Jian in Nantong at the end of the nineteenth century is considered the earliest practice in modern China. Some studies held that Zhang Jian initially conceived the regional planning idea of "the sequent formation of settlements, towns, and cities" in his article "On Emperor Shun Being an Industrial Politician" (1904), envisioning the development of Nantong not only from the angle of the city itself, but also sought the coordinated development of cities, towns, and rural areas [1]. The proposed Nantong-Taizhou-Yancheng economic zone and the plan to open the port of Wusong predate today's Suzhou-Wuxi-Changzhou Economic Zone by more than half a century. In practice, since 1895, he tried to lay the economic foundation of regional modernization with industrialization as the driver and through the development of industry, education, and public welfare, improved the people's overall quality by education, and then improved the ecological and human environment of the region through transportation, water conservancy engineering and

K. Wang, *National Spatial Planning in China*, Urban Sustainability,
https://doi.org/10.1007/978-981-97-7729-7_4

public welfare—all this reflected how Nantong's modernization process is systematic and thorough. The "villageism" advocated by Zhang Jian epitomizes his idea of pursuing and practicing local enterprises and regional planning. "Villageism" treats the local area as a part of the state, or a "point" on the "surface", figuratively, seeking its own path of development through local autonomy in the turbulent times and harsh social environment. Zhang Jian believes in "achieving self-governance through self-survival, self-reliance, and self-defense"[1] "Nantong, though a mere county, is as good as a village in China", and if Nantong is well-managed, then the areas adjacent to it are settled, and plans would be turned into reality.

Some studies believe that the realization of "villageism" in space is a point-to-surface process. Zhang Jian had intended to use the areas having access to the sea as the starting point to establish a new political and economic center in northern Jiangsu that was independent of Jiangning, with the aim of promote local autonomy on a larger scale and gradually realize the plan of "growing a province starting from Xuzhou". Although the plan failed, Zhang Jian continued his relentless efforts in developing beyond Tonghai township (Nantong-Haimen), and his ideal of "the sequent formation of villages, towns, and cities" is embodied in his attempts starting from the reclaimed and Kenmu village to northern Jiangsu, and then to "growing a province from Xuzhou". Zhang Jian took the whole process of regional development into consideration, reorganizing and transforming the region through planning and construction, with the goal of establishing a "prototype of a new world". Apart from treating planning as a link in regional development and a means to solve the problem of regional development, he had his mind on the bigger picture of the "overall improvement of society" [1].

The study argues that among his contemporaries like E. Howard (1850–1928) and P. Gaidis Geddes (1854–1932), the founders of the theories of modern urban planning and regional planning, Zhang Jian not only took a holistic approach to urban construction and regional development, but also put it into practice, hence the enduring significance of his accomplishment [1]

4.1.2 The Strategies for Founding a Country by Sun Yat-sen Laid the Foundation for National Spatial Planning

Dr. Sun Yat-sen, the forerunner of the Chinese revolution, put forward *The Strategies for Founding A Country*[2] in 1921, it which he conceived the first thoughts on the construction of the entire nation from many aspects. It can be regarded as the

[1] See Vol. 4, Enterprises, "Preface to the Report of the Autonomous Council", *Complete Works of Zhang Jian*, The Library of Nantong, ed. Nanjing: Jiangsu Ancient Books Publishing House, 1994. 465.

[2] Originally named "*The International Development of China*", it was written in English between 1918 and 1919 and first published in the June issue of the *Far Eastern Review* in 1919. In October 1921, Shanghai Minzhi Book Company published its Chinese translation.

earliest prototype of national spatial planning in China's modern history. In it, Dr. Sun proposed a series of major construction plans that still have an important impact on the development of many cities.

The *The Strategies for Founding A Country* comprises three parts: "psychological construction", "material construction", and "social construction", of which "material construction", also known as "industrial plan", served as a comprehensive planning concerning China's economic development and infrastructure construction. The plan proposed specific development ideas by the six distinct regions according to Sun Yat-sen's vision, covering transportation, housing, industry, and other aspects.

According to Sun, industrial development was the key to China's survival, and an important factor contributing to warlordism was that transportation in various places was extremely underdeveloped, economic ties were rather weak, and there was no unified market. Therefore, the "Industrial Plan" was mainly organized by region, and prioritized transportation construction. Its six major plans are: The first plan is to build the northwest railway system centered on major northern port. The second plan is to improve the waterway and banks of the Yangtze River with the major oriental port as the center. The third plan is to build the southwest railway system centered on the major southern port. The fourth plan is a detailed railway construction plan, that is to build five major railway systems in the central, southeastern, northeastern, expanded northwest, and the plateau areas, as well as to set up manufacturing plants of locomotives and passenger and freight cars. The fifth plan is targeted at industries concerning people's livelihood, including food, clothing, housing, movement, printing and so on. The sixth plan is about the mining industry, including mining for iron ore, coal, oil, copper, and special minerals, as well as the manufacturing of mining machinery and the setting up of smelter factories.

The "Industrial Plan" set the following ten goals covering transportation, commercial ports, cities, water conservancy, industry, mining, agriculture, irrigation, forestry, and relocation: (1) development of transportation; (2) opening of commercial ports; (3) new market streets opened in railway centers, the terminals and commercial ports, equipped with public facilities; (4) development of water conservancy; (5) set up large factories for smelting iron, steel and cement yard to meet the needs of realizing the above goals; (6) development of mining industry; (7) development of agriculture; (8) irrigation in Inner Mongolia and Xinjiang; (9) afforestation in northern and central China; (10) population relocation to the three northeastern provinces, Inner Mongolia, Xinjiang, Qinghai, and Xizang. To sum up, Sun sincerely hoped that all countries will join hands in tapping into China's rich resources through international economic cooperation to enhance productivity and improve people's living standards.

Some studies held that the "Industrial Plan" is a construction plan of unprecedented scale that draws on international experience. Sun Yat-sen had studied the works of renowned experts in planning: *Cities in Evolution* by P. Geddes (1854–1932) and *The Modern City and Its Problems* by F.C. Howe (1867–1940). In addition, Sun also read the latest works on urban planning published between 1912 and 1916, including *Modern City Planning and Maintenance* by F. Koester, *Replanning Small Cities* by J. Nolen, *Introduction to Urban Planning* by J. Jullan, *Town Planning in Practice* by R. Unwin and so on. The study of these works proved that Sun had drew extensively on the successful experience of Europe and the United States and had considerable planning expertise when writing the "Industrial Plan".

Although the "Industrial Plan" failed to materialize due to the semi-feudal and semi-colonial state of China and was a utopia at the time, it remained a work of lasting significance in that Dr. Sun made a detailed analysis of the country's economic development from the national perspective for the first time, and the spatial distribution and planning of a series of ports, railways and other transportation facilities especially played a crucial role in the construction of the spatial structure of the nation as a whole.

4.1.3 Regional Planning Practices of the then National Government

After its establishment in April 1927, the then National Government in Nanjing set out on establishing relevant institutions for economic development. In April 1935, the former Construction Committee, the National Economic Council, and the National Defense Commission were reorganized into the Resource Commission "for coordinated arrangement and with the task of developing national resources and handling matters related to industries including national defense and mining. Its priority work is the establishment of a hinterland defense economy" [2]. From April 1935 to March 1938, the scope of work of the Resource Commission shifted from investigation, research, and design in the early days to the establishment of industrial and mining enterprises in later time.

In March 1936, according to the *Outline of the Implementation Plan for National Economic Construction* adopted by the First Plenary Session of the Fifth Central Committee of the Kuomintang in 1935, the Resource Commission formulated the *Preliminary Plan of National Defense Industry*, the main content of which is to establish a nationalized heavy industry zone in Jiangxi and Hunan, and exploit the mineral resources of the southwestern provinces, and draw up the *Five-Year Plan for Heavy Industry Construction*, which was later revised into the more detailed *Three-Year Plan for China's Industrial Development*. These construction plans established the centrality of national defense and gave priority to the construction of heavy industry.

When the Anti-Japanese War broke out in 1937, the focus of the Resource Commission shifted to the study of the three-year plan for the defense industry, the five-year plan for post-war industrial construction, and the investigation and study of regional economy. During the war, the commission mainly focused on the development of heavy industry in central and western China, establishing 11 central industrial areas, thus changing the previous imbalanced industrial distribution of heavily concentrated in the eastern coastal areas. So, it was as early as a few years prior to the victory of the war that the commission began to consider post-war reconstruction. In 1943, Weng Wenhao published the article "The Problem of Industrialization in

Postwar China". Most workers in the industrial and mining enterprises under the Resource Commission returned to work in quite a short period of time after the victory, and many industrial and mining enterprises were taken over from the hands of Japanese invaders. As a result, the Resource Commission rapidly expanded to become the largest and most concentrated administration of China's heavy industry.

In the field of academic research, Ji Chaoding published his influential doctoral dissertation "Key Economic Areas in Chinese History, as Revealed in the Development of Public Works for Water-Control" in 1936, in which he categorized "economic areas" into "key economic area" and "subsidiary economic area". "Key economic area" refers to "the area where the conditions of agricultural production and transportation facilities are much superior to other areas as far as the provision of grain tribute in concerned, and whoever is able to control the area would likely to subdue and unify all of China…" He holds that economic areas are the economic basis of unity and division of China throughout the history and the geographical basis of the demarcation of local administrative regions. His study elevated the significance of regional analysis to a new height. Due to the outbreak of the War of Liberation, the KMT-led national government failed, and the industrialization, urban and regional planning in modern China also came to a halt.

4.2 Four Crucial Periods of Urban Spatial Adjustment in China in the Past Fifty Years

4.2.1 Impact of the 156 Projects[3] in the First Five-Year Plan on the Spatial Distribution of Cities and Towns

After the founding of New China in 1949 and with the successful completion of the restoration work in the following three years, China began to implement its *First Five-Year Plan* of the national economy in 1953, ushering in the largest industrialization process in Chinese history. A large-scale economic construction with 156 construction projects as the core and 900-odd large and medium-sized projects (projects above designated size) as the supplement, the *First Five-Year Plan* served as the cornerstone and milestone of China's industrialization [3]. Since the 156 projects and their supporting projects were implemented and distributed nationwide and taking urban construction into account from the outset, it has had an extremely important impact on the spatial distribution and development scale of cities across the country. In nature, it played the role of the first national plan of urban spatial distribution since the founding of New China.

[3] The *Industry Cornerstone of New China—Studies of the 156 Construction Projects* by Dong Zhikai and Wu Jiang held that out of the "156 projects" of China and the former Soviet Union, 154 were determined, of which 2 were counted twice and 2 were not built at all, so the true figure should be 150.

The *First Five-Year Plan* was formulated in accordance with the Party's general line and tasks during the transitional period. Its basic task was to devote all forces to the industrial construction consisting of the 156 construction projects with the aid of former Soviet Union and other projects above designated size. The plan stipulated that the total expenditure on economic and cultural construction within the five years will be 76.6 billion CNY, of which 42.7 billion (55.8% of the total expenditure) will be invested in capital construction. And of the 42.74 billion, industrial construction will account for 58.2%,[4] with priorities on metallurgy, coal, and machinery. Considering such factors as combat readiness and resources in determining spatial distribution, enterprises of iron and steel, non-ferrous metals, and chemical industries were in the central and western regions with abundant mineral resources and sufficient energy supply; the machinery industry was in areas near the raw material production base. The 150 projects constructed were distributed in 54 cities of 17 provinces, of which 106 were civil enterprises. Except for the 50 that were in the northeast region, the rest were in the central and western regions: 29 were in the central region, and 21 in the western region. Of the 44 defense enterprises, apart from some shipyards that were located along the coast, 35 were founded in the central and western regions. The development path prioritizing heavy industry adopted by the *First Five-Year Plan* has had an important impact on China's industrial structure, regional pattern, and the construction of new industrial cities. In 1956, the country announced that the plan was realized one year ahead of schedule. In 1957, China's gross national product, industrial and agricultural output value, and fiscal revenue all increased by more than 60 percent over 1952, marking remarkable achievements of the plan.

The success of the *First Five-Year Plan* is largely due to the general forward-looking spatial distribution of the 156 projects and the scientific distribution of specific projects. In the early days of New China, more than 77% of the country's total industrial output was concentrated along the long strip of area in the east (its area accounted for less than 12% of the country's total), 68% of which was achieved in a handful of places including the Yangtze River Delta centered on Shanghai (23%), the southeast centered on Shenyang (20%), the Beijing-Tianjin-Tangshan region centered on Tianjin (10%), Qingdao-Jinan Railway region centered on Qingdao (5%), and Guangzhou (3%). In the northwest and Inner Mongolia, which account for 45% of the country's land area, the total industrial output value accounted a mere 3% of the country. Sichuan, Yunnan, Guizhou, and Xizang with an area of the country's 23% contributed only 6% to the total industrial output [4]. The pattern was not conducive to the sound allocation of resources and to the country's economic security [3]. In 1956, Mao Zedong pointed out in his well-known speech "On the Ten Major Relations" that "China's industries are concentrated in the coastal areas, and by that, I mean Liaoning, Hebei, Beijing, eastern Henan, Shandong, Anhui, Jiangsu, Shanghai, Zhejiang, Fujian, Guangdong, and Guangxi. About 70 percent of all our country's light and heavy industries are in the coastal areas, leaving only 30 percent in the hinterland. This is an unacceptable situation developed over history[5]."

---

[4] Compiled according to the "Review of the Ten Five-Year Plans" on www.sohu.com.

[5] See "On the Ten Major Relations", *Selected Works of Mao Zedong*.

To support the spatial distribution and construction of the 156 key projects, the industrial ministries of the central government collected data of more than 200 cities and towns across the country and analyzed their conditions for setting up factories. In April 1953, Li Fuchun, director of the then State Planning Commission, led a "joint group for factory site-selection" composed of nearly 100 carders from various industrial ministries and departments of railways, health, water conservancy, electric power, public security, culture, urban construction and so on, technical personnel, and experts of former Soviet Union for on-site surveys in key candidate cities including Zhengzhou, Luoyang, Xi'an, Lanzhou and so on. Finally, 15 key cities including Beijing, Taiyuan, Xi'an, Zhengzhou, Lanzhou, Baotou, Wuhan, Chengdu, Shenyang, Jilin, Harbin, and Fularji were chose as the sites for 91 projects.

In September 1953, the CPC Central Committee pointed out "The planning of key industrial cities must be stepped up, and more efforts should be made to facilitate cities where a large proportion of industrial construction are carried out. Work concerning urban planning and design must be strengthened so that a draft urban master plan might be drawn up as quickly as possible and submitted to the central government for review[6]." More than 150 cities across the country prepared urban master plans, and the then National Construction Commission and urban development departments reviewed and approved the master plans of 15 cities where key industrial projects were located, such as Taiyuan, Lanzhou, Xi'an, Luoyang, and Baotou. In addition, the cities of the country were classified as follows to cater for the development of industrialization: Category I, 8 cities devoted to heavy industry: Beijing, Baotou, Xi'an, Datong, Qiqihar, Daye, Lanzhou, and Chengdu; Category II, reconstructed cities with a relatively large proportion of industry: 14 cities of Jilin, Anshan, Fushun, Benxi, Shenyang, Harbin, Taiyuan, Wuhan, Shijiazhuang, Handan, Zhengzhou, Luoyang, Zhanjiang, and Urumqi; Category III, old cities with a small proportion of industry: 17 cities of Tianjin, Tangshan, Dalian, Changchun, Jiamusi, Shanghai, Qingdao, Nanjing, Hangzhou, Jinan, Chongqing, Kunming, Neijiang, Guiyang, Guangzhou, Xiangtan and Xiangfan; Category IV, cities other than the 39 key cities above. Wanli (former Vice Premier of the State Council) later concluded, "Thanks to the foresight we'd had back then, we developed a number of key cities during the First Five-Year period, and they are still faring well today[7]."

In May 1956, the CPC Central Committee stated in the "Decision on Several Issues Concerning the Strengthening of the Construction of New Industrial Zones and New Industrial Cities" that "socialist construction requires sound allocation of the country's productive forces and an important step in achieving this is by actively conducting regional planning and making rational arrangement for the newly built industrial enterprises and human settlements during the second and third five-year plan periods[8]." The decision also elaborated on the significance of regional planning, pointing out that "regional planning is about carrying out comprehensive planning for

---

[6] See "Working Framework of Urban and Rural Planning under Market Economy in China". Ministry of Housing and Urban–Rural Development, 1997.

[7] See "Wanli on Urban Construction". China City Press, 1994. 284.

[8] Compiled based on the course materials of "Regional Planning" by Wu Tinghai.

such capital constructions as industry, power, transportation, post and telecommunications facilities, water conservancy, agriculture, forestry, human settlements, and construction bases and various engineering facilities in areas that are to be developed into new industrial zones and new industrial cities, in accordance with the specific natural and economic conditions of the localities as well as their long-term development plans of national economy. The objective is to ensure the sound development of the construction of new industrial zones and new industrial cities by making sure there is good coordination and cooperation between the components of national economy and among the industrial enterprises in a certain area, the spatial distribution of human settlements is reasonable, and the construction of various projects is orderly." In 1956, the regional planning for the following ten areas began: Baotou-Hohhot, Xi'an-Baoji, Lanzhou, Xining, Zhangye-Yumen, Sanmenxia, Xiangfan, central Hunan, Chengdu, and Kunming. Wan Li said on many occasions that "the 156 projects in the *First Five-Year Plan* period were carried out based on a unified plan, and the results were good."

The 156 projects were distributed in compliance with the following major principles: first, considering the security of national defense, they were first and foremost located in hinterland; second, they must be close to places of raw materials and fuels; third, existing urban living and production infrastructures should be made use of as much as possible; fourth, they must be near traffic arteries and transportation hubs; fifth, due consideration must be given to the development of ethnic minority areas and underdeveloped areas [5]. The results of implementation had the dual effects of both guaranteeing the security of national economy and coordinating the imbalance in the development of the eastern and western regions. 65% of the projects were in 45 cities and 61 workers' towns (towns built especially for the workers of the projects) to the west of the Beijing-Guangzhou Railway; 35% were in 46 cities and 55 workers' towns to the east of Beijing-Guangzhou Railway and in the northeast region. Meanwhile, several cities were chosen as the sites for centralized distribution of construction projects, so that these cities can achieve scale of production capacity to a certain degree in a short period of time, which not only promoted the development of industry, but also boosted the construction and development of cities. Although there was no clear national urban spatial planning during the construction of the 156 projects, its industry-oriented, city-based spatial distribution was like the early land resources planning in Japan and South Korea as well as the early national spatial planning in European countries such as the Netherlands. So, it can be said that the spatial distribution of 156 key projects during the *First Five-Year Plan* period played the role of the first national spatial planning of the New China.

4.2.2 The Disconnection Between "to Back Against Mountains, to Scatter, and to hide in Caves" and the Urban Development During the "Third Front Movement"

In 1964, the Central Committee and Mao Zedong, upon consideration of the serious international situation and the danger of war at the time, put forward the idea of strengthening the construction of the strategic "Third Front" and actively preparing for war. In April 1965, the Central Committee issued the "Instructions on Strengthening War Preparations", making the decision to speed up the construction of the strategic home front in all provinces and autonomous regions across the country. In accordance with the spirit of Mao Zedong's instruction, the formulation of the *Third Five-Year Plan* and the *Fourth Five-Year Plan* as well as production and construction all shifted to war preparation and the "Third Front Movement". The large-scale development and construction in the central and western regions during the "Third Front Movement" used more than 50% of the country's investment in capital construction, carried out 380 coastal projects, and relocated 145,000 workers and more than 38,000 pieces of equipment. In the whole construction process, "no concentrated cities were built", and all cities in the country felt the impact[9]. It can be considered the second large-scale adjustment of the spatial structure of New China, only the cost was high.

The "First Front" refers to the frontier areas of strategic importance, including the southeast coastal region and the northeast region. The "Third Front" refers to the national home front of strategic importance, to be specific, "the area that is west of the Beijing-Guangzhou Railway, east of Wushaoling in Gansu, and south of Yanmen Pass in Shanxi", including the three provinces of Sichuan, Guizhou and Yunnan in the southwest region, the four provinces of Shaanxi, Gansu, Ningxia, and Qinghai in the northwest region, the western parts of the three provinces of Henan, Hunan, and Hubei in the central and southern regions, the northern part of Guangdong Province, the northwest part of Guangxi Autonomous Region, and the western parts of the two provinces of Shanxi and Hebei in North China. The "Second Front" refers to the transition zone between the "First Front" and "Third Front" (Fig. 4.1).

The construction of the "Third Front" can be roughly divided into two periods[10]. During the *Third Five-Year Plan* period, the construction of the "Third Front" was focused on the southwest: several important railway trunk lines connecting the southwest were constructed, including Sichuan-Guizhou Railway, Chengdu-Kunming Railway, Guiyang-Kunming Railway, Xiangyang-Chongqing Railway, and Hunan-Guizhou Railway; two major iron and steel bases in Panzhihua and Jiuquan were constructed, and ten relocation and follow-up projects for national defense services carried out; in coal industry, the priority was to build 12 mining areas in Guizhou Province, including Liuzhi Special Region, Shuicheng County and Pan County; in power industry, the priority was the construction of hydro power stations such as

[9] See *Urban Construction in Contemporary China*. Beijing: China Social Science Press, 1990. 88.
[10] [4]

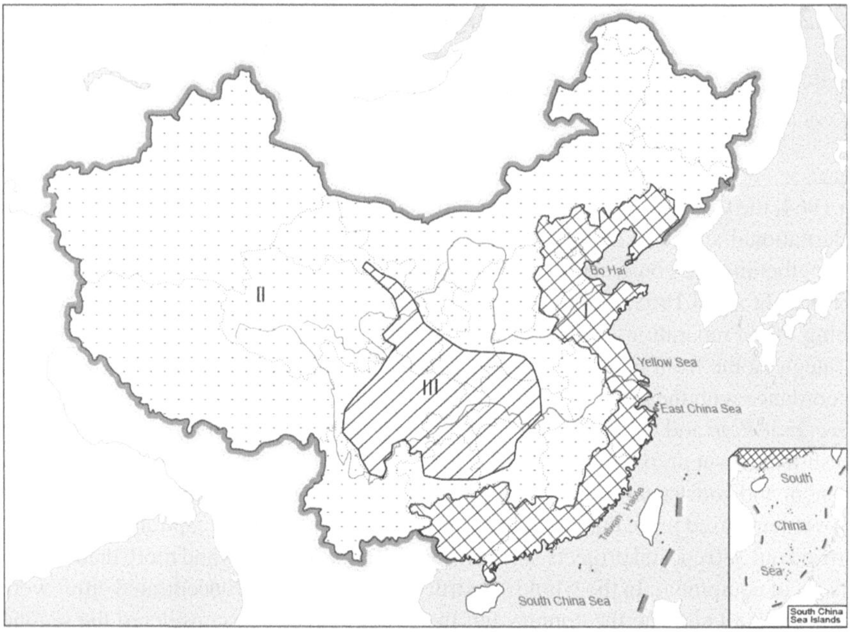

Fig. 4.1 Diagram of the three fronts. *Source* Regional Development Report of China, 1997, Lu Dadao et al. The Commercial Press, 1997

Yingxiuwan and Gongzui in Sichuan Province, and Liujiaxia in Gansu Province, and of thermal power stations such as Jiajiang in Sichuan and Qingshan in Hubei; in oil industry, the focus was on the development of natural gas in Sichuan; in machinery industry, the focus was on the construction of Sichuan Deyang Heavy Machinery Factory, Dongfeng Electric Machinery Factory, Guizhou Bearing Factory, etc.; in chemical industry, the focus was on projects that serve national defense. During the five years, construction investment in the country's mainland reached 63.1 billion CNY, accounting for 64.7% of the country's capital construction investment. Of it, investment in the 11 provinces and autonomous regions in the "Third Front" areas was 48.2 billion CNY, accounting for 52.7% of the total. During the *Fourth Five-Year Plan* period, the focus of "Third Front" construction was shifted to the western parts of the three provinces of Henan, Hubei, and Hunan while the construction of the Great Southwest continued. During this period, the country was divided into 10 economic cooperation zones, including Southwest, Northwest, Central Plains, South China, East China, North China, Northeast China and Shandong, Fujian and Jiangxi, and Xinjiang. And each zone was required to gradually establish the industrial system and national economic system that are of different levels, with its own characteristics, self-reliant, and highly cooperative (Shandong, Fujian and Jiangxi, and Xinjiang were required to build a "small yet complete" economic system), and each province (municipality, and autonomous region) was required to establish its own "sub-Third Front".

During the "Third Front" construction, the state relocated many old enterprises in coastal areas to the "Third Front" area in a planned manner. In 1965, the then State Construction Commission determined the relocation principles of "dispersion in general and small-scale concentration", and a few cutting-edge national defense projects need to be "to back against mountains, to scatter, and to hide", and some would even have to go into caves. In March 1966, the Southwest Bureau of the CPC Central Committee, in its meeting in Chengdu, Sichuan on the "Third Front" construction in the southwest, summarized the factory-building experience of taking factories as the main form and avoiding cities. After nearly ten years of construction, nearly 2000 large and medium-sized enterprises and scientific research units were set up in the "Third Front" area, 45 large-scale production and scientific research bases and 30 new industrial cities were founded, and a huge "Third Front" industry with more than 1/3 of the original value of industrial fixed assets of the country and devoted to national defense industry and mechanical and electrical industry was established. However, due to these factories were scattered and the artificial severance of the organic connection of production, both the macro and micro economic benefits dropped significantly compared with the *First Five-Year Plan* period. Worse still, the fact that the factories were intentionally removed from cities in all aspects brought considerable inconvenience to workers' life, which was in sharp contrast to the *First Five-Year Plan* period, when integration with the cities was always emphasized, and consideration was even given to the right proportion of male and female workers in different industries of a city.

Although the "Third Front" construction played a certain role in the founding of several industrial centers and new industrial cities[11] in certain areas of the western region, the losses caused by the construction were greater. First, it had seriously negative impact on the economic development of coastal cities. During the "Third Front" construction, funding was heavily invested in the western region, and the division and large-scale relocation of factories, design, and research institutes to the "Third Front" area seriously weakened the production capacity of the coastal cities. Second, due to the principles of "to back against mountains, to scatter, and to hide in caves" followed in site-selection of factories, the factories were cut off from the cities and all by themselves in terms of geographical location, which made it impossible for production capacity to grow or cities to form in a long time to come.

Entering the 1980s, many units had to relocate extensively for a second time, moving back to their original cities or nearby areas, incurring huge economic losses and serious social consequences. If only the "Third Front" construction had followed the industrial spatial distribution principle adopted by the *First Five-Year Plan*, we would have seen several modern cities with economic strength to promote the development of the central and western regions and facilitate balanced regional development. From a historical point of view, the "Third Front" construction represented another major turning point in China's industrial space and urban development. But

[11] For example, the iron and steel cities of Panzhihua, and Jiayuguan; the coal cities of Liupanshui, Tongchuan, and Shizuishan; the automobile city of Shiyan; the non-ferrous metal cities of Jinchang, Baiyin, etc.

the mistake caused by the wrong guiding ideology cost us the opportunities for the adjustment and development of the national urban spatial structure and for the second national spatial planning.

4.2.3 Strategy of Opening the Coastal Areas and Development of Coastal Urban Belts in the 1980s

In 1978, the development strategy of Reform and Opening Up was adopted by the Third Plenary Session of the 11th CPC Central Committee, and the coastal areas attached importance of unprecedented level, heralding yet another major turning point in the spatial structure of China's cities and towns. In 1979, when formulating the *Sixth Five-Year Plan*, the Central Committee, upon reflection on and stocktaking of the lessons learned from the distribution of China's productive forces over the past 30 years since the founding of the People's Republic, stated clearly in the *Sixth Five-Year Plan* for the national economic and social development that it is imperative to make active use of the current foundation of the coastal areas and "give full play to their special strengths to further promote the economic development of the hinterland." During the *Sixth Five-Year Plan* period, the distribution of China's productive forces was aimed at improving economic efficiency and tilted toward areas with preferable conditions. Of the investment in national capital construction, the shared received by coastal areas increased from 42.2% during the *Fifth Five-Year Plan* period to 47.7%, while that of the hinterland decreased from 50.0% to 46.5%. The change in investment structure reflects the shift of the country's development priorities[12].

Beginning in 1978, our country once again pursued an unbalanced development strategy. From 1979 to 1980, China set up four special economic zones in Shenzhen, Zhuhai, Shantou, and Xiamen. In 1984, the 14 port cities of Dalian, Qinhuangdao, Tianjin, Qingdao, Yantai, Lianyungang, Nantong, Shanghai, Ningbo, Wenzhou, Fuzhou, Guangzhou, Zhanjiang, and Beihai along the coast were opened and national economic and technological development zones were established. In 1985, the Yangtze River Delta, Pearl River Delta and Hokkien Golden Triangle were designated as economic open zones. In 1988, Hainan Province in its entirety was approved as special economic zone. In 1990, the decision on developing Pudong New Area was made. All the above marked yet another major shift of the focus of China's industrial space and urban development to the coastal area. Since 1980, the coastal area has seen dramatic economic growth in which development zones played the leading role. According to statistics, the 29 national coastal development zones generated a GDP of 452.4 billion CNY in 2004, accounting for 2.5% of the country's total, achieved an industrial added value of 332.8 billion CNY, accounting for 4.4% of the country's total, and utilized a total of 9.2 billion USD of foreign

[12] [4]

investment, accounting for 15.2% of the country's total. Several "star cities" distinguished themselves by their performance, including Shenzhen, Zhuhai, Dalian, and Qingdao (Table 4.1).

The construction of development zones has important impact on the urban development of coastal areas. On the one hand, the scale of existing cities was greatly expanded; cities such as Shanghai, Dalian, Qingdao, and Ningbo have more than doubled in terms of population and land use in the past 20 years. On the other, many new cities emerged, the outstanding representatives of which are Shenzhen and Zhuhai, and many small and medium-sized cities were flourishing in the Yangtze River Delta and the Pearl River Delta, such as Nanhai, Shunde, Zhangjiagang, Kunshan, Xiaoshan and so on. More importantly, the coastal cities have gradually developed into the core area buttressing China's participation in global competition. The town concentrated areas centering on the Yangtze River Delta, the Pearl River Delta and the Beijing-Tianjin-Hebei region have become the main bases of export-oriented economy, and the form of metropolitan interlocking region began to show, marking a major shift of China's urban form from the development of individual city to the development of urban cluster. This efficient and developed form of urban area is of great significance to both the urbanization progress and modernization of towns in China (Fig. 4.2).

From the perspective of planning, the expansion of coastal urban space is in essence the growing of development zones. With the coordination and unification of the construction and management of the development zone and the urban area, the development of coastal cities is itself a process of continuous expansion of industrial land and growing urban population; urban development is just like the growing of a giant development zone. The policy of development zones lay the foundation for the overall pattern of urban space at the macro level in China in the past 20 years and contributed to the general pattern of descending order in terms of urban economy, scale of cities and towns, and urban development quality from the eastern region to the central and western regions. Although the resulting regional gap is much larger than imagined, it laid the general foundation of China's urban space in the next 20 years.

At the time when the coastal areas were developing by leaps and bounds, the state started the work related to land development, improvement, and planning since 1981 [4]. After the Third Plenary Session of the 11th CPC Central Committee, a leading cadre of a certain province, upon returning from a visit to three Western European countries, proposed the following in his report to the Central Committee: "There are many experiences in regional land development that we can learn from the Western European countries…… in particular, while they are doing it, they would first set a goal, then coordinate actions in all aspects aiming at it and make sure the measures are implemented. This way, the results are often good. This is something we should learn[13]." Since then, the leading cadres of the CPC Central Committee and the State Council have repeatedly asked the then National Construction Commission to push ahead with the work of regional development. Many far-sighted people in theoretical

[13] [6]

Table 4.1 List of development areas/zones/district approved by planning review before 1995

Province/municipality	Development areas/zones/ district	Time of approval	Area approved (ha)
Beijing	Beijing economic-technological development area	1994.8	3980
Tianjin	Tianjin economic-technological development area	1984.12	3300
Hebei	Qinhuangdao economic and technological development zone	1984.10	690
Liaoning	Shenyang economic-technological development zone Yingkou economic and technological development zone Dalian economic and technological development zone	1993.4 1992.10 1984.9	1000 560 2000
Jilin	Changchun economic and technological development zone	1993.4	1000
Heilongjiang	Harbin economic-technological development zone	1993.4	1000
Shanghai	Minhang economic and technological development zone Hongqiao economic and technological development zone Caohejing Hi-Tech park	1986.1 1986.8 1988.6	308 65.2 1330
Jiangsu	Kunshan economic and technological development zone Suzhou industrial park Nantong economic and technological development area Lianyungang economic and technological development zone	1992.8 1994.2 1984.12 1984.12	1000 7000 2429 300
Zhejiang	Hangzhou economic and technological development zone Xiaoshan economic and technological development zone Ningbo economic and technological development zone Wenzhou economic and technological development zone	1993.4 1993.9 1984.10 1992.3	1000 920 2960 511
Anhui	Wuhu economic and technological development zone	1993.4	1000
Fujian	Fuzhou economic and technological development zone Fuqing Rongqiao economic and technological development zone Dongshan economic and technological development zone	1985.1 1992.10 1993.1	1000 1000 1000

(continued)

Table 4.1 (continued)

Province/municipality	Development areas/zones/ district	Time of approval	Area approved (ha)
Shandong	Qingdao economic and technological development zone Yantai economic and technological development area Weihai economic and technological development zone	1984.1 1984.10 1992.1	1250 1000 572
Hubei	Wuhan economic and technological development zone	1993.4	1000
Guangdong	Guangzhou economic and technological development district Guangzhou nansha economic and technological development zone Huizhou daya bay economic and technological development zone Zhanjiang economic and technological development zone	1984.12 1993.05 1993.05 1984.11	3857.72 2760 998 920
Hainan	Yangpu economic development zone	1992.3	3000
Chongqing	Chongqing economic and technological development zone	1993.4	960
Xinjiang	Urumqi economic and technological development zone	1994.8	430
Total	33		45,100.92

Source China Development Zones Year book 2005

and academic circles also said on many occasions that China should carry out land use planning. For example, in June 1980, Yu Guangyuan proposed to conduct research on the land resources economics at the National Forestry Conference. This could be called the stage where the public were prepared for China's national land work.

In April 1981, the 97th meeting of the Secretariat of the CPC Central Committee issued instructions on carrying out land improvement work: "The National Construction Commission should cooperate with the State Agricultural Commission in carrying out national land improvement. The Construction Commission should not limit its task to infrastructure projects, but also include land use, land development, comprehensive development, regional development, environmental improvement, and development of large rivers. It is imperative to promulgate laws and make plans. Land improvement is a big problem, and many countries have special ministries devoted to this matter. Instead of establishing a separate ministry, we could set up a special agency under the National Construction Commission to decide on the tasks and plans for the State Council to review and approve. In short, we must act and

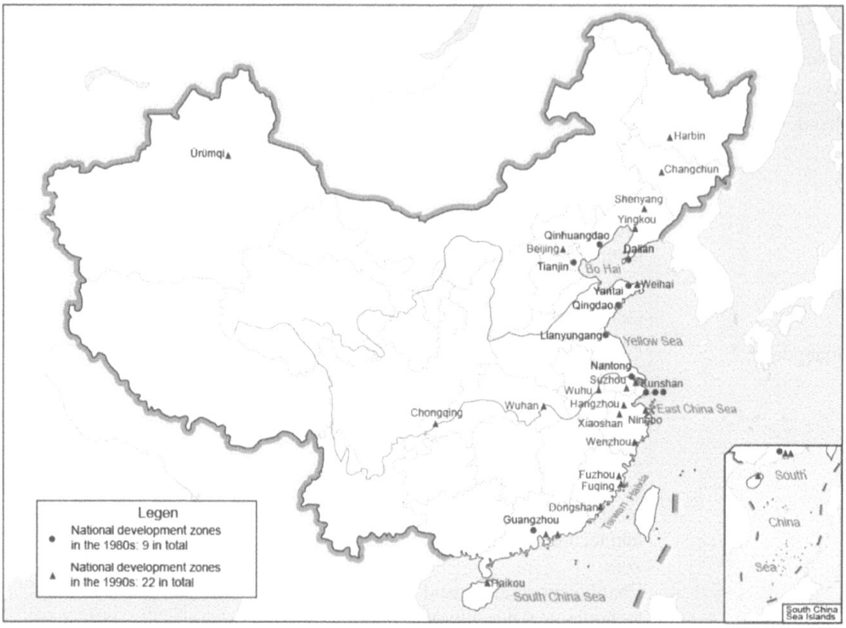

Fig. 4.2 Spatial distribution of national economic and technological development zones before 1995 (drawn by author)

manage our land well" (Bi Weiming, 1983). On August 15 the same year, the National Construction Commission submitted the Report on Carrying out Land Improvement to the State Council. Cadres of the State Council quickly gave instructions: Start the work first and improve it step by step in the process of doing it. On October 7, the State Council officially forwarded the report to the departments of all provinces, municipalities and autonomous regions and pointed out in the instructions: "In a big country like ours, it is a very important task to manage our land well. At present, the land resources and ecological balance of our country have been seriously damaged, and a lot needs to be done in the development and utilization of land resources, hence the urgent need to strengthen the work of land improvement. Since the work covers a wide range of areas, it is hoped that all localities and departments will work closely together to do a good job in this matter of great importance." In addition, it is pointed out that land improvement includes work in the following five aspects: investigation, development, utilization, governance, and protection, and the Land Bureau tasked with responsibilities concerning land work was approved to be set up under the former State Capital Construction Commission [7].

The State Council carried out institutional reform at the beginning of 1982, abolishing the National Construction Commission, and placed the Land Bureau under the leadership of the State Planning Commission. Since 1984, the State Planning Commission organized to have the land resources planning formulated with the participation of relevant ministries and commissions of the State Council.

The Ministry of Housing and Urban–Rural Development (formerly the Ministry of Urban–Rural Construction and Environmental Protection), as a participant in the planning, put forward a systematic idea for the development of urban space across the country by drafting the National Urban Spatial Development Strategy 2000 (more on this later).

Beginning in 1982, the State Planning Commission carried out a succession of pilot projects in the following areas: Beijing-Tianjin-Tangshan, Songhua Lake in Jilin, Yichang in Hubei, coastal area in Ningbo, Zhejiang, Bayingol Mongolian Autonomous Prefecture in Xinjiang, and western Henan Province. In July and September 1982, the northern and southern pilot experience-exchanging and on-site inspection meetings were held in Songhua Lake and Yichang, respectively; and in 1983, it was fully rolled out based on further pilot projects in various provinces, autonomous regions, and municipalities. To better guide and promote the work of land resources planning, formulation of the *Outline of the Overall Land Resources Planning* began in 1984 after several years of exploration and practice, and the *Measures for Land Resources Planning Formulation* (the *Measures*) was promulgated in 1987. The *Measures* stipulates the nature, role, tasks, contents, principles of formulation, and approval procedures for land resources planning. According to it, by nature land resources planning is "an important part of the national economic and social development planning system, a guiding plan for the comprehensive development of resources, the overall spatial distribution of construction, and comprehensive environmental improvement, and an important basis for the preparation of medium and long-term plans." According to the *Measures*, the task of land resources planning is: "Coordinate well the following relationships from the perspective of the area in its entirety and by taking account of the advantages and characteristics of the areas under planning: the relationship between the development and utilization of land resources and its governance and protection, and the relationship between population, resources, and environment, to promote the comprehensive development of the regional economy."

The *Outline of the Overall Land Resources Planning* published in 1987 proposed the general framework for China's land development and spatial distribution of productivity with coastal areas and the areas along the Yangtze River and the Yellow River spanning the east and west as the main axes, and other transportation arteries as the secondary axes, and identified 19 key areas for comprehensive development in future (Table 3.3). Subsequently, several cross-provincial (autonomous regions and municipalities) land resources planning was carried out in succession, including the Panzhihua-Xichang-Liupanshui area, the border area between Hunan, Jiangxi and Guangdong, the border area between Shanxi, Shaanxi and Inner Mongolia, the lower reaches of the Jinsha River, and the areas along the mainstream of the Wujiang River. Hainan Province also organized the formulation of the *Comprehensive Development Plan for Hainan Island* (1988), a joint effort of China and Japan. By the early 1990s, 22 provinces, municipalities, cities under separate state planning, 223 prefectures (cities, prefectures), and 640 counties had conducted land resources planning (Bi Weiming 1994) (Table 4.2).

Table 4.2 19 Key areas of land development

	Scope	Area (10,000 km2)	Population (10,000)	Advantages of regional development	Main factors restricting regional development
1. Beijing-Tianjin-Tangshan area	Beijing, Tianjin, and Tangshan, Qinhuangdao and Langfang in Hebei Province	5.5	2672.1	Geographically important location, convenient transportation, and strong economic foundation; abundant mineral resources such as iron ore, coal, and so on, and great potential of oil reserves	Serious shortage of water resources, industries excessively concentrated in the three major urban areas of Beijing, Tianjin and Tangshan, irrational urban structure, and deterioration of environmental quality in certain areas
2. Yangtze River Delta	10 cities of Shanghai, Suzhou, Wuxi, Changzhou, Nantong, Hangzhou, Jiaxing, Huzhou, Ningbo and Shaoxing, and 55 counties	7.5	5153.6	Ideally located with convenient water and land transportation; sound agricultural foundation with the reputation of the "land of fish and rice"; strong industrial foundation with complete categories and high economic benefits; scientifically and technologically developed, and abundant intellectual resources; densely distributed cities and high level of urbanization; developed domestic and foreign trade, large area of tidal flats along the coast, and abundant tourism resources	Severe energy shortages, raw material shortages, insufficient transportation, and infrastructure capacity; excessively concentrated industries, and serious environmental pollution
3. Central and southern Liaoning	10 cities of Shenyang, Dalian, Fushun, Anshan, Benxi, Liaoyang, Tieling, Dandong, Yingkou, Panjin, and 30 counties	7.7	2271.5	Abundant resources such as iron ore, magnesite, boron ore and so on, strong foundation of heavy industry, and a complete range of categories	Insufficient energy, scarce water supply, and polluted environment

(continued)

Table 4.2 (continued)

	Scope	Area (10,000 km2)	Population (10,000)	Advantages of regional development	Main factors restricting regional development
4. Pearl River Delta	Guangzhou, Foshan, Jiangmen, Shenzhen, Zhuhai, Dongguan	3.7	1510.8	Adjacent to Hong Kong and Macao, convenient water, and land transportation; relatively complete categories of industries with good foundation; the main production areas of subtropical cash crops; and good prospects of petroleum and petrochemical industry	Lack of energy sources
5. Shandong Peninsula	Qingdao, Yantai, Weifang, Weihai, Dongying, Rizhao	6.0	2156.1	Convenient sea transportation made possible by multiple ports; rich in minerals such as oil, gold, graphite and so on; and well-developed tourism and marine fishing industries	Severely scarce water resources
6. Golden Triangle of Southern Fujian	Xiamen, Zhangzhou, Quanzhou and their 9 counties including Tong'an, Jinjiang, Hui'an, Nan'an, Anxi, Yongchun, Dongshan, Longhai and Zhangpu	1.4	1017.0	Rich in subtropical fruits and cash crops; many natural harbors, abundant aquatic resources; and superior geographical location, which is conducive to the development of export-oriented economy	Weak foundation of transport and urban infrastructure, and the investment climate needs to be improved

(continued)

Table 4.2 (continued)

	Scope	Area (10,000 km2)	Population (10,000)	Advantages of regional development	Main factors restricting regional development
7. Hainan Island	Hainan Island	3.4	600.0	China's largest special economic zone, rich in tropical resources with great land potential, numerous harbors, good prospects for marine development, and rich tourism resources	Shortage of energy, backward cities and transport facilities, and insufficient funding
8. Hongshui River Hydropower & mineral development zone	Includes both sides of the river from the Tiansheng Bridge of the Nanpan River to the Datengxia of the Qianjiang River, covering parts of Guangxi and Guizhouu	3.6	499.1	Abundant water resources and non-ferrous metal mineral resources	Weak economic foundation, lack of funds and talents, and insufficient transportation
9. Energy development areas in Yanzhou, Tengzhou, Xuzhou, and Huainan&Huaibei	Spans the four provinces of Jiangsu, Shandong, Anhui and Henan, covering the cities and counties of Jining, Zaozhuang, Xuzhou, Huainan, Huaibei, Bengbu, Fuyang and Su, etc	4.6	2600.0	Abundant coal reserves of good quality and complete varieties, and great potential for agricultural development	Need to strengthen the construction of water supply facilities in traffic river and step up efforts in disaster prevention and environmental protection
10. Harbin-Changchun area	Harbin, Qiqihar, Daqing, Changchun, Jilin	6.9	1629.4	Rich in natural resources such as petroleum and hydropower, large land area and large per capita arable land, economic foundation to a certain level	Energy and transportation infrastructures need to be improved

(continued)

Table 4.2 (continued)

	Scope	Area (10,000 km2)	Population (10,000)	Advantages of regional development	Main factors restricting regional development
11. Energy base centered on Shanxi	Shanxi and parts of Shaanxi and Inner Mongolia adjacent to it, western Henan, and Dawukou of Ningxia	23.1	4721.2	China's largest energy base, vast coal reserves, good quality and complete varieties, superior mining conditions, considerably rich in bauxite, rare earths, molybdenum, and copper, relatively good industrial foundation	Few external transportation routes, insufficient transportation capacity, lack of water, weak agricultural foundation, and fragile ecological environment
12. Coastal area of the middle reaches of the Yangtze River centered on Wuhan	Includes the coastal area from Yueyang, Hunan to Jiujiang, Jiangxi	4.3	1966.1	Convenient water and land transportation, rich in such mineral resources as copper, iron and gypsum, strong industrial foundation, and great potential for agricultural development	More efforts need to put in the comprehensive improvement of the Yangtze River and flood control
13. Coastal area along Yangtze River from Chongqing to Yichang	Coastal area along Yangtze River from Chongqing to Yichang	3.6	1202.2	Rich in hydropower, phosphorus, iron ore and natural gas resources, great tourism development value of the Three Gorges, and strong industrial foundation of Chongqing	Need to strengthen the construction of an integrated transportation network combining railways, highways, and water transport
14. Border area of Hunan, Jiangxi and Guangdong	Chenzhou in Hunan, Ganzhou in Jiangxi, and Shaoguan in Guangdong	8.0	1370.0	rich in mineral resources such as tungsten, antimony, lead, and zinc, and in biological resources	Weak economic foundation and backward transportation
15. Hydropower and non-ferrous metallurgical areas in the upper reaches of the Yellow River centered on Lanzhou	Coastal areas along the mainstream of the Yellow River from Longyang Gorge in Qinghai to Qingtongxia in Ningxia, and Jinchuan and Changba in Gansu where non-ferrous metal resources are distributed	5.0	484.1	Particularly abundant hydropower resources and non-ferrous metal mineral resources, and the resources are well matched	Fragile ecology, serious soil erosion, lack of funds, technology and talents for resource development, and the Baotou-Lanzhou Railway needs to be renovated asap

(continued)

Table 4.2 (continued)

	Scope	Area (10,000 km2)	Population (10,000)	Advantages of regional development	Main factors restricting regional development
16. Area along the mainstream of the Wujiang River	Coastal area from Guizhou to Pengshui, Sichuan	4.0		Rich in water energy and mineral resources such as aluminum and phosphorus	Inconvenient transportation, and serious soil erosion upstream
17. Panxi-Liupanshui Development Zone	Includes parts of Panzhihua, Yibin, Liangshan Yi Autonomous Prefecture in Sichuan, parts of Liupanshui and Bijie in Guizhou, and Zhaotong in Yunnan	8.4	1477.8	Good heavy industry foundation, and rich in vanadium, titanium, phosphorite, coal, non-ferrous metals and hydroenergy resources	Remotely located, inconvenient transportation, lack infrastructure and supporting facilities, and serious soil erosion
18. Karamay, Urumqi	Cities of Urumqi, Shihezi, and Karamay	5.0	223.9	Abundant oil, coal, limestone resources, vast land, and great potential of solar and thermal resources	Scarce water resources and infrastructure need to be strengthened
19. Hydropower and non-ferrous metal base in the middle reaches of the Lancang River	Includes Dali Bai Autonomous Prefecture, Nujiang Lisu Autonomous Prefecture, Baoshan District, Lincang Region, Simao District	4.7	375.0	Rich in mineral resources such as water energy, lead, zinc, tin, and antimony, and in animal and plant resources and tourism resources, and superior conditions in terms of light, thermal and water energy	Backward transportation

Source Land Improvement and Economic Construction, Bi Weiming. Capital Normal University Press, 1994

The practice of land resources planning in the 10 years from the 1980s to the 1990s played a positive role in promoting the macro-level scientific use and management of national and regional space resources in a comprehensive manner. In the practice of land resources planning, the distribution of productive forces and urban development based on economic zoning have contributed to the comprehensive coordination of regions. However, the focus of land resources planning is on the use of national land resources, therefore, the understanding of the role of urban space in promoting social and economic development was insufficient. Although there was the spatial distribution plan for the development of coastal cities and towns, it wanted system and integrity, and there was a lack of a thorough regional development policy. As a result, the impact it generated was far less than the five national comprehensive plans of Japan and the Netherlands' national spatial planning.

4.2.4 Regional Spatial Pattern Centered on Coastal Area and Diversification Since 2000

Since 2000, with the establishment of the socialist market economy and the decentralization of power from the central administration to local governments, economic development has become the priority for all the regional governments. Moreover, with the country's increasing reliance on export-oriented economy, the connection between cities and between cities and regions has been strengthened. Various research and planning at the regional level occurred as a result. All in all, the development of cities and towns in the coastal areas is much faster than that in the central and western regions. In terms of urbanization level, the fifth census (China's fifth national economic census) showed that the levels of urbanization of the three major zones in the eastern, central, and western regions are 44.60%, 33.50% and 27.66%, respectively, with a difference of nearly 17 percentage points between the east and the west, and the urbanization level of developed provinces and regions such as Guangdong, Jiangsu, and Zhejiang have exceeded 50%. The scale of the central cities in the eastern region kept growing, and several urban clusters and metropolitan coordinating regions were gradually forming. Although little importance was attached to the national research in urban space during this period because it was considered a product of the planned economy, an active interest was shown in the spatial research and planning at the regional level, especially in the eastern region. A few regional spatial planning of large, medium, and small-scales emerged since 2000, such as the large-scale regional spatial planning of Beijing-Tianjin-Hebei region and the Pearl River Delta, the medium-scale planning of the metropolitan coordinating region or urban clusters of Suzhou-Wuxi-Changzhou region and Hangzhou Bay, and the small-scale research in urban spatial development strategy concerning such cities as Guangzhou and Ningbo. The emergence of these practices of different types and scales reflects the following reality: the Chinese cities must move towards a greater space for development under the circumstance of deepening globalization and marketization. And the governments of cities must also take a regional approach to development in the new situation. The practice of various regional spatial planning plays a crucial role in promoting the establishment of urban planning concepts based on regional coordinated development.

(1) Macro-level Research on Regional Urban Space

Urban and Rural Spatial Development Planning Study for the Beijing-Tianjin-Hebei Region

The "Urban and Rural Spatial Development Planning Study for the Beijing-Tianjin-Hebei Region" led by Professor Wu Liangyong of Tsinghua University and jointly completed by 10-odd units in 2001 is the most important regional spatial planning research in recent years.

According to its concept, the Beijing-Tianjin-Hebei region comprises two triangular regions covering the cities of Beijing, Tianjin, Tangshan, Baoding, Langfang, etc., with a population of about 40 million (in year 2000) and a land area of 70,000 km2. It is roughly the ancient capital and its adjacent area in history.

From a global perspective and considering the latest fruits of international regional spatial planning research, the study put forward the development strategy of the Greater Beijing Area for the sake of national interests, and stated clearly for the first time that the Beijing-Tianjin-Hebei region should put in active efforts in building itself into one of the world's major metropolitan regions in the twenty-first century, laying the foundation for gaining national competitive advantages. In terms of spatial structure, the study proposed to realize the strategic vision of "dual-core, multi-center metropolitan coordinating region" by relieving central cities of non-essential functions on one hand and "re-concentration" in certain regional space on the other. With Beijing and Tianjin as the dual cores of the main axis and Tangshan and Baoding as the wings, it aimed at relieving large cities of non-essential functions, boosting the development of medium-sized cities, and increasing urban density to foster the urban cluster of the Greater Beijing. As for the development of regional towns, the development model of "transportation axis, 'grape bunches', and ecological greening" was proposed. In terms of regional coordination, the study proposed to establish an inter-administrative division—the Planning and Construction Commission of the Greater Beijing Region—for the coordination and cooperation in major cross-regional projects (such as transportation, ecology, environment, and industrial structure).

Although the study is aimed at the Beijing area, it is quite macro in the analysis method and planning concept, and open and dynamic in the spatial structure. And thanks to the orientation made clear by its vision of "dual-core, multi-center metropolitan coordinating region", the study played a positive role in promoting subsequent research on the urban space development strategy of Beijing and Tianjin and the formulation of a new round of overall urban planning of the two cities.

Planning of the Pearl River Delta Urban Cluster

The "Urban Cluster Planning of the Pearl River Delta" developed by the China Academy of Urban Planning and Design in 2004 is an operable regional spatial planning responding to the particularities of economically developed areas. The Pearl River Delta comprises the seven cities of Guangzhou, Shenzhen, Zhuhai, Foshan, Dongguan, Zhongshan and Jiangmen, the Duanzhou District and Dinghu District of Zhaoqing City, Gaoyao City, Sihui City, and the Huicheng District, Huiyang District, Huidong County and Boluo County of Huizhou City, covering a land area of 41,698 km2, accounting for 23.20% of the province's total. It is a region with sustained and rapid economic growth since China's Reform and Opening Up.

Guided by the strategic goal of building a moderately prosperous society in an all-round way, the planning aimed at promoting regional economic and social development of a greater area, strengthening international economic cooperation of the region, and consolidating its competitive position. It was further proposed that the Pearl River Delta build itself into a global manufacturing base and a world-class urban cluster and develop an economic region with comprehensive advantages and functions covering such industries as processing and production, modern logistics, financial and professional services, tourism and entertainment, and information center. It also clarified the ecological "bottom line" and ecological system framework for regional development. In terms of spatial structure, it was proposed that in future, a highly integrated, networked, and open regional spatial structure and urban functional distribution system will be formed in the delta. Through the spatial structure of "one ridge, three belts and five axes", the most important functional areas and nodes of the delta would be connected and integrated to form a spatial system that radiates to the open sea and inland area in eight directions. By constructing the "dual-core, multi-center and multi-level" central system and the "three metropolitan areas with diversified development", sub-regional urban space systems with its own characteristics and a network of regional and local service centers will be formed.

In terms of space control in development and construction, the planning proposed the new approach of four-tier control: supervision-oriented, regulation-oriented, coordination-oriented, and guidance-oriented. Supervision-oriented control was applied to areas of regional green spaces and major traffic corridors, thus subject to mandatory supervision and control by the provincial government through legislative and administrative means. Regulation-oriented control was applied to areas where basic industries and heavy equipment manufacturing of the region are located and to area where major transportation hubs are located, which was governed by the provincial governments through such means as planning, guidance, arbitration and so on, and governments of specific cities are responsible for the actual development and construction. Coordination-oriented control was aimed at the cross-border cooperation and development of Guangdong-Hong Kong-Macao area and the coordinating area of inter-city planning and construction, which is subject to control by the provincial governments by means of planning, guidance, consultation, etc. Guidance-oriented control was applied to economic revitalization and supporting areas, urban development and upgrading areas, and other policy areas in general. The provincial governments would provide development guidance on the type and scale of development, ecological environment requirements and construction standards of the region, and offer support for development policies and infrastructure construction. Governments of specific cities will develop independently in accordance with the guidance.

For regions and areas with important regional significance and influence, major action plans of the provincial and municipal governments which are jointly participated by relevant departments will be put forward in accordance with the objectives of planning and development. Practical planning guidelines concerning policy zoning, hierarchical spatial governance and urban spatial coordination will be proposed according to the characteristics of regional planning and the requirements of planning implementation.

The planning of the urban cluster in the Pearl River Delta is most prominently characterized by the fact that it proposed capacity needs of space and controlled area of ecological environment in response to the continued rapid development and puts forward a feasible plan for the implementation of the plan, which is of pioneering significance (Fig. 4.3).

(2) Spatial Planning Research of Urban Clusters

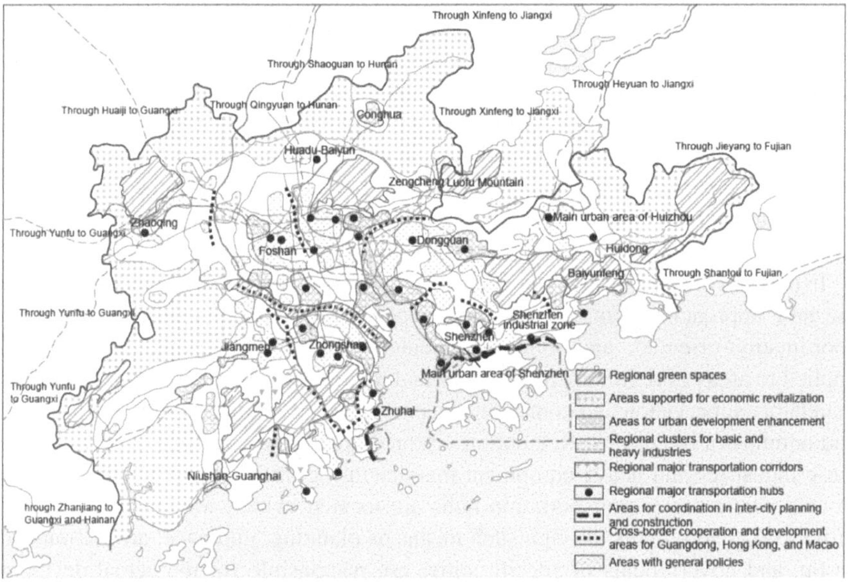

Fig. 4.3 Diagram of the policy zoning of pearl river delta. *Source* "Urban Cluster Planning of Pearl River Delta", China Academy of Urban Planning and Design, 2004

Suzhou-Wuxi-Changzhou Urban Agglomeration Planning

With a combined area of 17,500 km2 and a population of 13,542,100, the Suzhou-Wuxi-Changzhou region is the most economically developed region in Jiangsu Province, and the area receiving the most radiation effects of Shanghai. The Suzhou-Wuxi-Changzhou urban agglomeration planning is largely designed to seek common economic development by connecting with Shanghai.

In terms of regional functional positioning, Suzhou-Wuxi-Changzhou is the leading area and innovation center of economic development in Jiangsu, an organic part of the Shanghai metropolitan coordinating region, and the most competitive modern manufacturing base in the country and even the whole Asia–Pacific region. The plan placed the focus on the development of the secondary industry, intending to build the region into a sophisticated manufacturing center, a high-tech innovation base, and the main theater of economic internationalization. In terms of spatial organization, the spatial form of metropolitan area that combines "compact city and open area" was proposed. To be specific, cities and towns should be concentrated and compact to an appropriate extent, and independent and scattered towns and villages should be guided to gather step by step mainly by the development of a few key spatial areas and the matching of large-scale infrastructures, thus forming a dense urban belt with both high-level economy and high-quality space.

The plan laid down requirements for the rural space that has been neglected in the past: promote the concentration of urban and rural settlements in an active, orderly and sensible manner through such means as the hierarchical allocation of infrastructure and public service facilities, create conditions for the concentration and development of key central towns, and actively guide the concentration of scattered residential areas and the towns and villages that have been withdrawn and merged in a sensible manner. The plan is characterized by the joint efforts the three cities put in space utilization and resource protection, the special attention to the construction of villages and towns, and the priority given to the withdrawal, merging and concentration of villages and towns (Fig. 4.4).

Planning of Nanjing Metropolitan Area

Nanjing, the capital of Jiangsu Province, has an important influence in central Jiangsu and northern Anhui. The scope of the planning includes all the administrative areas of the six cities of Nanjing, Zhenjiang, Yangzhou, Ma'anshan, Chuzhou and Wuhu, Xuyi County and Jinhu County of Huai'an City, and the urban area, He County and Hanshan County of Chaohu City. The planning sought to re-understand Nanjing in the context of the Yangtze River Delta with Shanghai as the core, and breaking through the administrative divisions, it tried to take a holistic approach to the development of urban areas in a larger scope (including Anhui).

The planning proposed to promote the integrated economic and market development of the metropolitan coordinating region and build an international sophisticated manufacturing cluster along the Yangtze River and an urban belt featuring the elements of history, culture, nature, and cities. It put forward the idea of developing an industrial cluster to attract international manufacturing industry. In terms

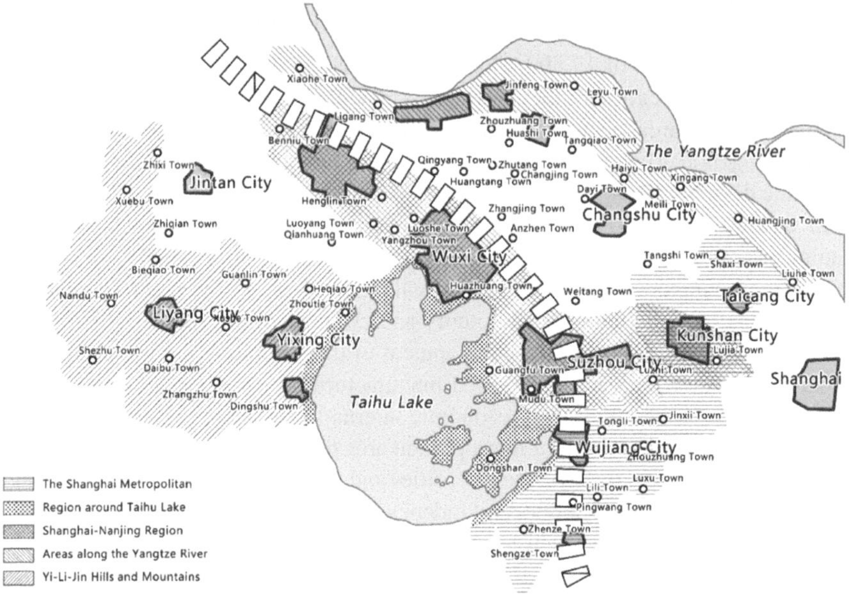

Fig. 4.4 Suzhou-Wuxi-Changzhou metropolitan area planning. *Source* Suzhou-Wuxi-Changzhou Metropolitan Area Planning, Jiangsu Urban and Rural Planning and Design Institute, 2001

of spatial structure, the form "one core and two circles" was proposed, focusing on the development of "one belt, one axis and three channels". To be specific, "one core" refers to the towns and potential urban development areas with the main city of Nanjing as the core and within a radius of about 30 km; and "two circles" refer to the "core circle" including the core city and the area within a radius of about 50 km from the center of the core city and the "close circle" within a radius of about 100 km from the center of the core city.

The planning proposed spatial zoning, requiring the areas along the Yangtze River to focus on the development of sophisticate manufacturing and modern service industries, preserving natural scenic spots, forest parks and protective green belts, guiding non-agricultural industries to relocate to the areas along the Yangtze River, and taking advantage of the construction of the urban development axes of Shanghai-Nanjing and Ningtong in the province, building highly developed urbanized area characterized by a high concentration of manufacturing and service industries and an organic combination of urban space and ecological space. The most important features of the planning are it breaks the administrative division, determines the scope of the metropolitan coordinating region based on the scope of economic radiation effects of the central city, and considers the metropolitan area as a part of the Shanghai metropolitan coordinating region (Fig. 4.5).

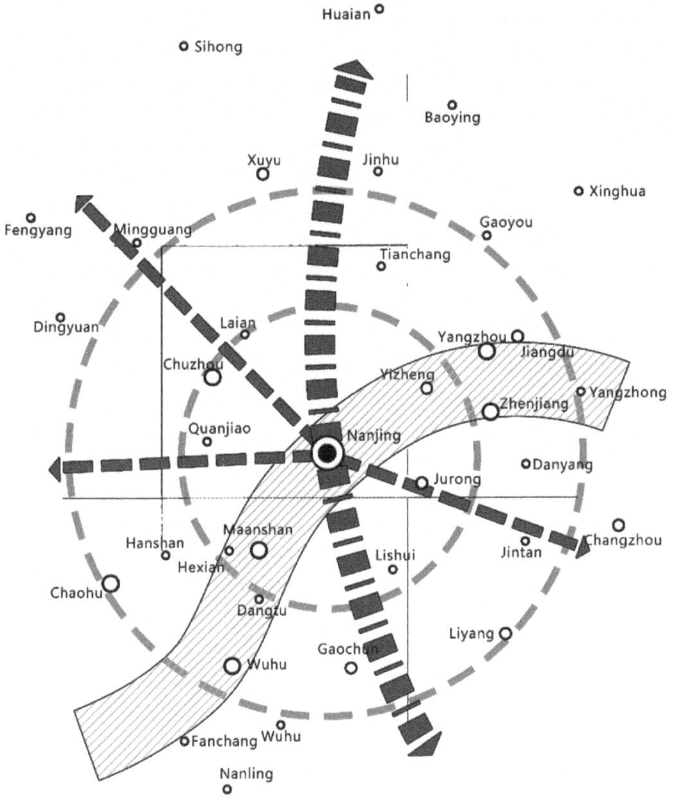

Fig. 4.5 Urban spatial organization of Nanjing metropolitan area. *Source* Planning of Nanjing Metropolitan Area, Jiangsu Urban and Rural Planning and Design Institute, 2001

Planning of Hangzhou Bay Urban Cluster

Adjacent to Shanghai, the Hangzhou Bay urban cluster is the economic center of Zhejiang Province. The planning of the cluster treats the areas under the jurisdiction of the six cities around Hangzhou Bay as a whole, which made it an exploration with constructive significance.

The urban cluster covers the six cities of Hangzhou, Ningbo, Shaoxing, Jiaxing, Huzhou and Zhoushan, with a population of 23.04 million in 2002 and a land area of nearly 45,400 km2. Guided by the idea of "connecting with Shanghai and taking an active part in cooperation and exchanges in the Yangtze River Delta region" proposed by Zhejiang Provincial Party Committee, the planning aims at building the area into the leader of new industrialization in Zhejiang and enhancing regional competitiveness by cultivating manufacturing bases and modern commerce and logistics industries that play a part in the world division of labor, advancing internationalization and promoting inter-regional division of labor and cooperation within the country. The planning also proposes to build the area as an integral part of the

Yangtze River Delta international metropolitan belt with Shanghai as the center, rendering it within three-hour traffic to Shanghai. In terms of space, the overall framework of "three belts, three clusters, four reserves, and six" was proposed, that is, to build three urban belts connecting to Shanghai, namely the belt along the Shanghai-Hangzhou-Ningbo Expressway, the coastal urban belt, and the belt around South Taihu Lake; foster the three major urban clusters and industrial concentration areas of Jiaxing-Huzhou, Hangzhou-Shaoxing, and Ningbo-Zhoushan; regulate the four types of major ecological reserves; cultivate six metropolitan areas of Hangzhou, Shaoxing, Ningbo, Zhoushan, Jiaxing and Huzhou that is in a continuous "V"-shaped distribution (Fig. 4.6).

According to the development plan of the Hangzhou Bay Industrial Belt, it is suggested that the overall scale of new planned industrial zone should be within 2000 km2 and maintain a 1:1 ratio with the planned urban space of central cities at all levels in the Hangzhou Bay area, and that the master planning of cities at all levels should take the industrial zones into consideration.

The planning is most prominently characterized by the construction of an industrial development belt along the coastal zone with Hangzhou Bay as the core. It can easily be seen from the map that the land in the coastal zone has basically been used up. In fact, Zhejiang Province, especially the Hangzhou Bay area, has seen severe water and energy shortages and serious pollution problems during its great industrial

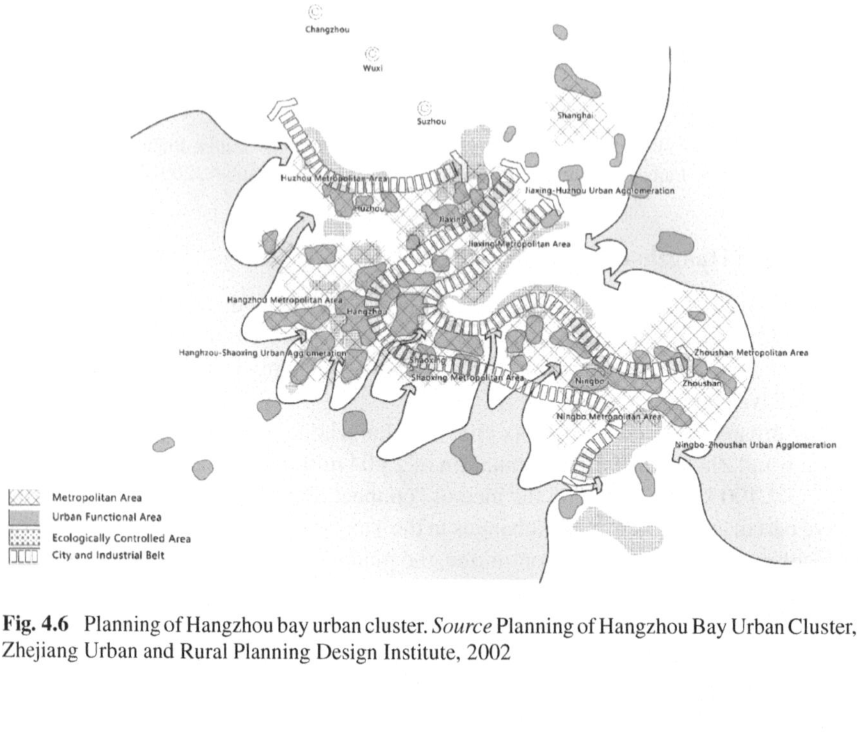

Fig. 4.6 Planning of Hangzhou bay urban cluster. *Source* Planning of Hangzhou Bay Urban Cluster, Zhejiang Urban and Rural Planning Design Institute, 2002

development over the past 20 years. The guiding principle and practices of the planning deviated from the concept of sustainable development to a considerable extent and did not meet the actual needs of local development.

All in all, the planning practice of different types of urban clusters is of pioneering significance. Although some plans failed to break the restrictions of administrative divisions, the conscious shift of focus from individual cities to coordinated development of regional towns represented an innovation in the concept of planning.

(3) Research on Urban Space Development Strategies

With the socialist market economy taking root and the increasing openness of the country since the 1990s, the original working methods and contents of overall urban planning were no longer compatible with the changing reality. As a result, local governments became keen on studies of spatial strategy and conceptual planning that is flexible, diverse, and forward-thinking. Cities include Guangzhou, Nanjing, Ningbo, Hangzhou, Beijing, and Tianjin have all launched studies of strategic planning with the hope of making up for the shortcomings of the general plan.

In fact, there is a broader and deeper context for the emergence of research on urban strategic planning. First, economic globalization requires cities, especially large cities, to adopt a global perspective. Second, the progress of urban regionalization and regional urbanization also require cities and regions to "stick together". As a new understanding in the academic circle goes: the urban area with central cities as the core is the basic spatial unit of urban competition in the era of globalization [8]. Therefore, the macro background of strategic planning is, in essence, the need of the socialist market economy when it reached a certain stage of development, and a comprehensive planning for the development of cities themselves in response to the rapid changes and severe challenges in the era of globalization [9].

Research on the Overall Development Strategy of Guangzhou

The overall urban development strategy of Guangzhou is the first of its kind as far as urban development strategy in concerned. Guangzhou is the central city of the Pearl River Delta and has been the political, economic, and cultural center of South China for years. With the global economic integration and the rapid development of the Pearl River Delta urban clusters and coastal open cities, Guangzhou is facing unprecedented challenges posed by Hong Kong and Shenzhen, and its status and role in the region declined accordingly. Meanwhile, given the problems including overlapping functions of the central urban area, "spillover" development, various functions competing for space resources, deteriorating environment, and significantly reduced urban efficiency, Guangzhou carried out the research of urban spatial strategy tentatively with the aim of promoting adjustment of urban spatial structure. In 2000, the former cities of Huadu and Panyu were reduced to districts, which increased the area of Guangzhou from 1443 to 7434 km2, thus creating conditions for the adjustment of the spatial structure of the central city.

Based on the growth of population and GDP, the plan puts forward the theory and ideas of Guangzhou's "leapfrogging" development by using inductive method to deduce the model of urban spatial growth. Judging by the future development trend of the Pearl River Delta region and Hong Kong, the plan determined the "dual-core model" of the Pearl River Delta metropolitan coordinating region and reconstructed the urban spatial structure of the Guangzhou, that is, to build a new Guangzhou with a population of 2.5 million in Nansha, Panyu, establishing a new "regional service center" in the Pearl River Delta region in future. At the same time, a new ecological development model is proposed, that is, a clustered urban layout studded with green space, and using the TOD development model to build a rail transit network in the Pearl River Delta region. This represents a bold innovation in the adjustment of urban spatial structure at the regional level. And building "new cities" has also become a major method adopted by many later planning of urban spatial development.

Guangzhou's urban spatial strategy of "leapfrogging" development is innovative, and it offers a new way of analyzing the development of large cities from the regional perspective. But the significance of the study is more theoretical than practical. Although the city eventually materialized the strategy of expanding towards the south and built the new city in Panyu, the result was far from the desired effect of a regional service center (Fig. 4.7).

Urban Development Strategy of Ningbo

A major economic city in Zhejiang Province, Ningbo is one of the central cities in the Yangtze River Delta. The city's urban space strategy was studied and developed thanks to the construction of the Hangzhou Bay Bridge (Ningbo Bridge), which has greatly changed the location condition of Ningbo, rendering it within three-hour traffic to Shanghai. The construction of Shanghai's international shipping center

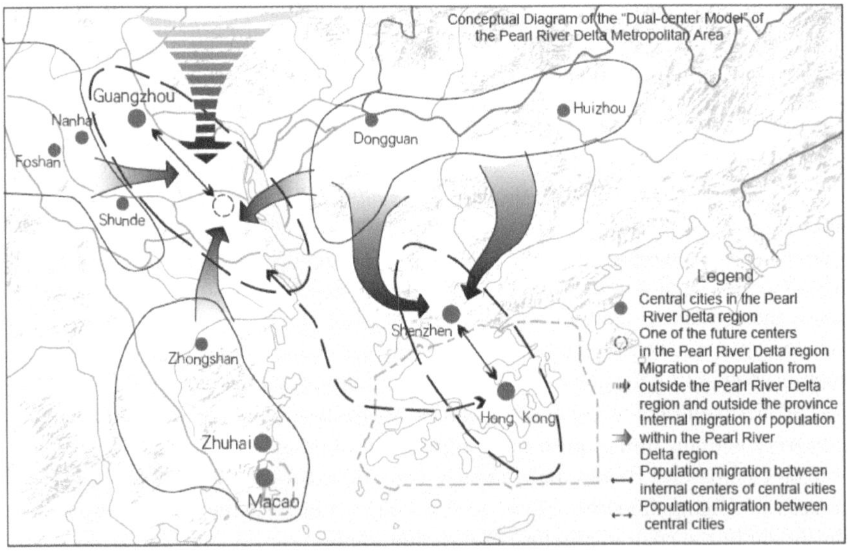

Fig. 4.7 Diagram of the "Dual-Core" structure of the Pearl River Delta. *Source* "Research on the Overall Development Strategy of Guangzhou", China Academy of Urban Planning and Design, 2001

(Yangshan Deepwater Ports) shook the status of Ningbo Port, and the large-scale construction of Shanghai Petrochemical Company in Jinshan District too had a significant impact on the structure of Ningbo's heavy chemical industry. Meanwhile, the inward-evolving type of urban spatial structure has outgrown the spatial framework determined 20 years ago, and it has difficulty adapting to the economic development in the 20 years to come.

In guiding ideology, the plan proposes to focus on improving the comprehensive competitiveness of the city, analyze its conditions in advantageous resources, especially special resources, and stress the role of market power in the port and industrial structure. In terms of spatial planning, it is proposed to reconstruct the "open" urban spatial structure embracing Hangzhou Bay, strengthen the central city, and build the framework of megacity by "expanding to the east and west, and build the north into working areas and the south into living quarter". In addition, the sub-center of the city will be built in the Yuyao-Cixi area by seizing the opportunity brought by the construction of the main corridor. In the analysis of the spatial structure of the central city, it is proposed to rediscover the "port-city relationship" and determine the relatively independent development model of Beilun Port.

The research on Ningbo's urban space strategy stressed the study on unexpected factors of urban development, focused on the analysis of the impact of market power in the development of industry and port, and constructed a structure embracing the greater Shanghai, which provides a good guide for the development of the entire city (Fig. 4.8).

Research on the Urban Space Development Strategy of Beijing

In 2003, Beijing carried out the research on urban spatial development strategy, which has generated some influence since it was initiated by the capital of China. Judging by its practical effect, the research, building on the fruits of the "Urban and Rural Spatial Development Planning Study for Beijing-Tianjin-Hebei Region", laid the foundation for the preparation of the overall urban planning of new Beijing. Since the objectives set in the previous version of the urban master plan (1993) were achieved ten years ahead of schedule and given the ever-increasing urban ills such as traffic congestion and environmental pollution, and the city's arriving at a turning point of development brought by the 2008 Summer Olympics, the research proposed the objectives of urban development and spatial structure of new Beijing.

The research estimates the future population size and construction land of the city by taking account of factors hampering urban development such as land resources, water resources, and green space, and proposed the objective of "building the capital into a world city with cultural charms and livable environment" and the general planning principle of "organic decentralization of old downtown area, migration within the municipal area, reintegration of villages and towns, and coordinated regional development". In terms of spatial structure, it is proposed to construct an urban spatial structure of "two axes, two belts and multiple centers" embracing the Beijing-Tianjin-Hebei region. The "two belts" at the macro level refer to the urban axis that connects the whole region, especially Tianjin, and at the micro level, it refers to the construction of several new cities within the municipal area to accommodate the

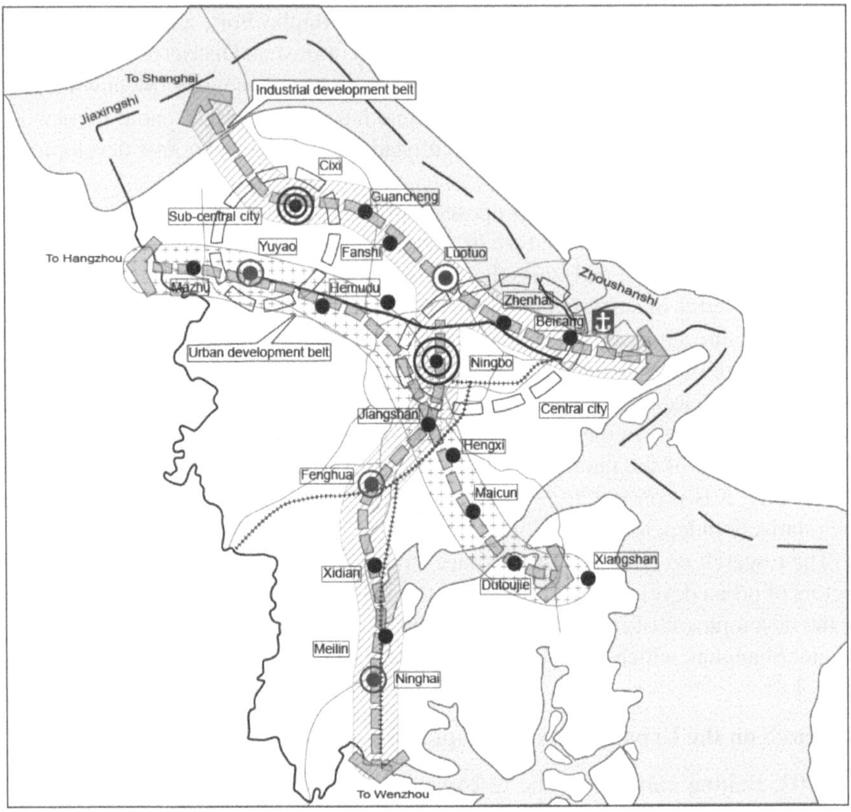

Fig. 4.8 Diagram of urban spatial structure of Ningbo. *Source* "Ningbo Urban Development Strategy", China Academy of Urban Planning and Design, 2001

city's new industries and population and host the surplus functions and population of the central area of the city.

As far as the analytical approach is concerned, the research clearly adopts the regional perspective, constructing the spatial structure of Beijing in the context of the entire Beijing-Tianjin-Hebei region, which is open-minded. In addition, the research's incorporating the constraints of resources and environment on urban development in its analysis is valuable in sensibly determining the scale of urban development (Fig. 4.9).

Even though it is not a thorough urban plan and lacks the breadth of regional planning and the depth of master planning, the practice of this planning has had an important impact on the theory, practice, and system of urban planning [8]. It demonstrates that urban planning has shifted from the study of development of individual cities to that of the entire region. As Mr. Zhou Ganzhi pointed out in the preface of *Introduction to Sciences of Human Settlements*, "the contents of urban planning have exceeded the scope of the past…The method of 'urban development strategy'

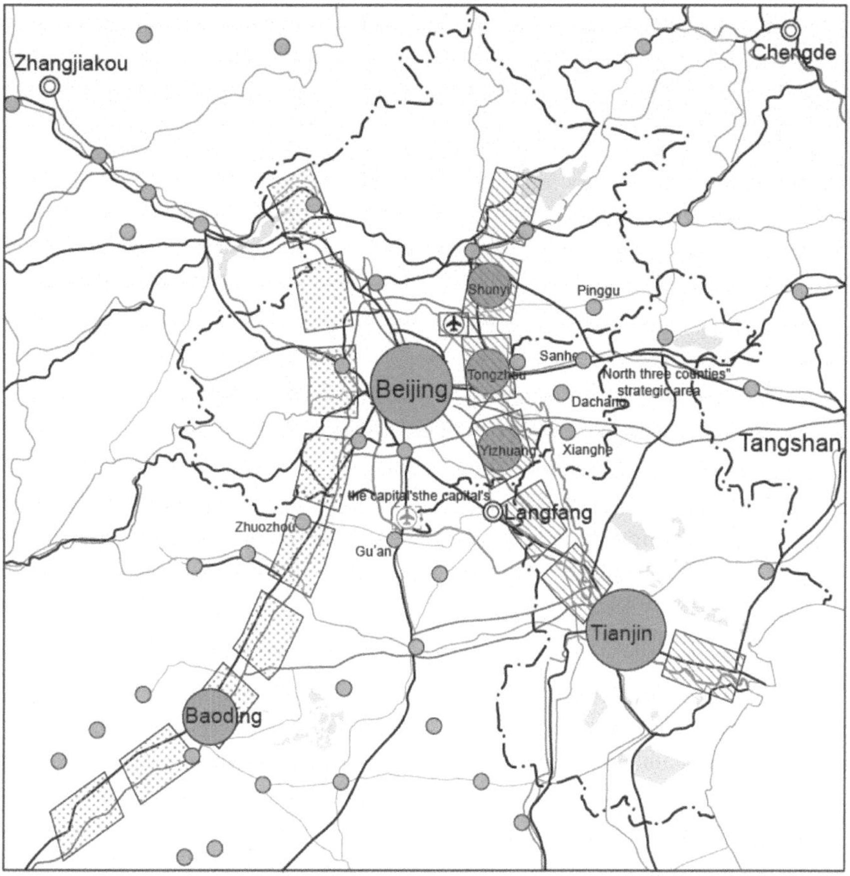

Fig. 4.9 Diagram of the urban spatial structure of Beijing. *Source* "Research on Urban Space Development Strategy of Beijing", Beijing Municipal Commission of Planning, 2003

has been widely adopted by large and medium-sized cities as an indispensable part of the job."

4.3 Review of China's Practices of Urban System Planning

According to the *Urban Planning Law* (1989), urban system planning is a statutory plan for the national, provincial, and municipal governments to organize urban space and guide the orderly and healthy development of cities and towns at different levels. The preparation of the national urban system plan is the main charge of the former Ministry of Construction, the urban planning administrative department of the State Council, and it plays an important guiding role in the development of cities and towns across the country and cross-regional coordination. Since the preparation of the national land resources planning by the then State Planning Commission in the 1980s, the former Ministry of Construction has organized many related studies and planning in this regard. Although the plans were not formally submitted for approval, they played a crucial role in the development of the whole country, especially in the formulation of national plans for a series of specialized sectors during the *Ninth* and *Tenth Five-Year Plan* periods. Therefore, a study of the three large-scale national urban system planning in 1985, 1999 and 2005 is helpful for us to gain a deeper understanding of the significance and role of national urban spatial planning.

4.3.1 Development Strategy of National Urban Spatial Distribution of 2000 Formulated in 1985

The preparation of the *Key Points of National Urban Spatial Distribution Development Strategy 2000* first started in early 1985 and was completed in October 1985. A special plan under land resources planning, it is the responsibility of the then State Planning Commission and led by the then Department of Urban and Rural Construction and Environmental Protection (predecessor of the former Ministry of Construction).

The main tasks put forward in the plan are to implement the major project plans determined by the state across the regions, to materialize the large-scale construction locations, to maintain a balance between the spatial distribution of cities and towns, productive forces, and population, and to promote the development of small towns. The plan is of strategic importance as it requires to clarify the key cities' nature, service scope, resource conditions, and development orientation and include them in the plan by taking account of the conditions of the cities themselves and the actual economic relations.

The main problems to be solved by the planning at the time were: (1) Weak horizontal connection across cities nationwide, a lack of sound division of labor between cities, many cities were developing all by themselves and their economic structures were similar, and the role of central cities was not fully played. (2) For cities along the coast and inland rivers, urban development was not fast enough, there were few cities, and the superior location advantage of being along the coast and rivers was not fully tapped into. (3) The spatial distribution of industries was illogical as many of them being far away from the cities and traffic lines, factories were scattered and distant from each other, resulting in huge waste. (4) The design of transportation system was not compatible with the layout of cities. In areas where cities were developing rapidly, road and rail construction was lagging. (5) The population of large cities was exploding. From 1979 to 1983, the net population increase of 48 large cities was 1.86 times that of small and medium-sized cities in the same period. In some cities, infrastructure was inadequate, there was a shortage of urban construction land, and environmental conditions were further deteriorating. (6) The imbalance of resources usage was aggravated, and rapidly developed industries and cities were excessively concentrated in the coastal and water-scarce areas in the north.

Based on the tasks and problems identified above, the following guidelines and principles of urban spatial distribution were established: (1) Integrate the layout of cities and productive forces, especially with the layout of industrial transportation construction projects, and make synchronized coordination as necessary. (2) Effectively implement the policy of "controlling the scale of large cities, developing medium-sized cities in a measured way, and actively developing small cities". Handle the relationship between urban and rural areas properly to create conditions for the gradual integration of urban and rural areas. (3) Handle the relationship between the eastern, central, and western zones properly, and establish a rational urban spatial distribution system step by step. (4) Give full play to the multi-functional role of cities, promoting horizontal economic ties between cities, and gradually building Chinese cities into open, multi-functional and civilized economic centers.

The objectives of urban development were set as follows: in 2000, the urban population will be reaching 360 to 400 million, accounting for 30 to 33.3% of the national population, and the number of cities will exceed 600 (the number of megacities will increase from 20 to 34, large cities will increase from 30 to 57, medium-sized cities will increase from 81 to 156, and small cities will increase from 169 to 397; the number of formally established towns will be 15,000 to 20,000).

The spatial distribution of cities and towns envisaged is as follows: (1) to form a multi-level urban system of different scales and different functions with central cities at all levels as the core and covering large, medium, and small cities. The first level comprises the central cities of national and international importance (Beijing, Shanghai, Hong Kong), the second level refers to the central cities of inter-provincial regions (Guangzhou, Wuhan, Chongqing, Tianjin, Shenyang, Dalian, Xi'an, Lanzhou), the third level includes the provincial central cities (about 35), the fourth level are the central cities of the provincial economic zones, and the fifth level refers to the central cities of counties. (2) The coastal, central, and western regions differ from each other in terms of the objectives of economic development in this century, therefore their urban spatial distribution and development policies should vary accordingly. (3) Continue to implement the policy of "controlling the scale of large cities, developing medium-sized cities in a measured way, and actively developing small cities", and categorize cities across the country into the following four types considering their specific conditions and provide targeted policy guidance: first, cities subject to strict control of sizes; second, cities subject to controlled development; third, cities whose development is encourage; fourth, cities subject to key protection.

The plan also puts forward relevant suggestions for implementation: (1) Establish the economic management system through reform and give full play to the role of central cities. (2) Promulgate supporting policies and measures, such as establishing towns or cities in newly built industrial and mining areas; offer small towns preferential treatment in terms of taxation, credit, energy supply, land use and social welfare; expand the scope of the household registration policy concerning non-agricultural population who take care of their own rations to county cities; and charge land use

Main problems the planning aimed to solve	Principles of planning	Development objectives for 2000	Layout of development	Measures to be taken
Illogical division of functions; insufficient development of coastal and riverside cities; misalignment between transportation and urban development; the population of large cities is expanding too fast; in some cities, urban development is not compatible with the resource conditions.	Coordinate urban development with the distribution of productive forces; control the scale of large cities and actively develop small towns; coordinate the development of the eastern, the central and the western regions; give full play to the multi-functional role of the cities.	Urban population: 360 to 400 million; urbanization level: 30 to 33.3%; cities and county-level cities, etc.: more than 600; formally established towns: 15,000 to 20,000.	Establish a five-level urban system; each of the three zones has its own priorities on urban development; provide targeted policy guidance of development and construction for the four categories of cities: those under strict controlled, subject to controlled development, whose development are encouraged, and those under key protection.	Facilitate central cities to play their role; support the development of small cities; charge land use fees; consider urban spatial distribution in site selection of key projects; establish an integrated transport network in the coastal zone; rational use of water resources.

Fig. 4.10 Summary of the key points of the national urban spatial distribution, 2000. *Source* Based on the Materials Provided by Zhang Qin of the Planning Division, former Ministry of Construction, 2005

fees to guide the rational development and spatial distribution of cities. Charges in large cities should be higher than in small and medium-sized cities, and in cities, charges in downtown areas should be higher than in suburbs. (3) The site selection of key projects should be compatible with the urban spatial organization—urban planning departments at all levels should participate in the preliminary work of site selection, and for key projects, planning departments should be consulted on site selection. (4) Build a comprehensive transportation network in the coastal zone as soon as possible. (5) Make overall arrangement for the planning and rational use of water resources (Fig. 4.10).

Afterwards, the Planning Division took stock of the implementation of the plan and existing problems, and although formal reporting and approval formalities lacked, the plan still played a role to certain degree. In 1998, China's urbanization level was 30.4%, with 668 cities and 18,800 towns. The spatial distribution of cities was adjusted, the cities' central function was enhanced, an urban network centering on the central cities at all levels was taking the initial form, and the areas along the rivers, coast, and transportation arteries became the areas witnessing the most rapid urban development in China. In formulating their own long-term development plans of the industry, the Ministries of Transport and Railways, and the State Oceanic Administration, etc. all took the *Key Points* as an important reference document. For instance, the Ministry of Transport decided to build the first and second-level central cities into main hubs of the highways and expressways, and former Ministry of Railways, in the planning of the high-speed public rail transit system, also took

Fig. 4.11 Spatial distribution of chinese cities and towns, *Source Development Strategy of National Urban Spatial Distribution 2000*, Planning Division, former Ministry of Construction, 1985

the first and second-level central cities as the starting points. Small cities have also seen unprecedented growth (Fig. 4.11)[14].

4.3.2 Preliminary Studies of National Urban System Planning

(1) Launch of the National Urban System Planning in 1994

In March 1994, the Urban Planning Division officially submitted a report to the leadership of the former Ministry of Construction for approval, recommending initiating the planning of national urban system. This was the first time that planning was carried out in the name of national urban system planning. The report stated that "the 14th National Congress of the CPC established the basic framework of the socialist market economy and further clarified the goals and arrangements for China's socialist modernization. Therefore, the time is ripe for the study of urbanization and spatial development strategies at the turn of the century. To this end, we plan to organize the preparation of a national urban system planning in accordance with the law and

[14] Based on the information provided by Zhang Qin of the Planning Division, Ministry of Urban and Rural Development.

submit relevant report to the State Council for review and approval. The resulting plan will serve as the guide to the formulation and revision of the urban master plan and an important basis for the state's decision-making on urban construction and development[15]." The report also held that there is foundation solid enough for carrying out this work, and several provinces and municipalities have already done some work on urban system planning which they planned to complete in two years. The head of the division expressed their support for work in their instruction, believing that "the planning of the (national) urban system is of great importance", and they further submitted the instruction to the State Planning Commission for approval[16]. It is the socialist market economy system determined by the 14th National Congress that made urban planning authority realize the necessity of sorting out the thinking on the development of the country's cities and towns in the new historical period and take this opportunity to establish a new urban spatial system.

(2) Research on Urban Development of the Areas along Longhai and Lan-Xin Railway

A key scientific research project of the former Ministry of Construction in the *Eighth Five-Year Plan* period, "Research on Urban Development of the Areas along the Longhai and Lan-Xin Railway" is an exploration of regional urban spatial distributionon a grand scale and can be regarded as the preliminary study of national urban system planning. Since the study spans the eastern, central, and western regions geographically, involving 25.48% of the country's land area, it provides theoretical and practical support for the development of national urban system planning.

The areas along the Longhai and Lan-Xin Railway refer to the long and narrow region that traverses the east and west of China from Lianyungang and Rizhao in the east, to Alashankou and Khorgos in the west, connecting the 10 provinces and autonomous regions of Jiangsu, Shandong, Anhui, Henan, Shanxi, Shaanxi, Gansu, Ningxia, Qinghai, and Xinjiang, with an area of 24,400 km2. The origin of the study can be traced to September 1990 when China's North Xinjiang Railway was connected to the Turkistan-Siberia Railway of Kazakhstan at Alashankou, forming the world's second Eurasian land bridge from Lianyungang and Rizhao in China in the east to the port of Rotterdam in the Netherlands in the west. Through analysis of the other four Eurasian land bridges in the world, the study made the preliminary conclusion that the economic development and urban construction of the areas along the Longhai and Lan-Xin Railway will usher in a new stage of development.

According to the study, the areas have been heavily invested by the state for decades and have become the base of China's heavy chemical industries including coal, oil, natural gas, non-ferrous metals and so on. But due to their industrial structure largely catering to heavy and chemical industries, the areas failed to fit in with local industries. The dual economic structure was formed because of this "embedded"

[15] Quoted from the report dated March 26, 1994, by the Urban Planning Division, the former Ministry of Construction.

[16] Ibid.

investment method, hence the failure to fully tap into the potential of the local areas and to effectively promote the common development of the entire region.

Based on the economic relevance, transportation convenience, similarities in history and administrative division of the whole region, the study divided the areas into eight urban clusters of the Huaihai, Central Plains, central and southern Shanxi, Guanzhong Plain, Ningxia, Gansu, Qinghai, and both sides of Tianshan Mountains in Xinjiang. Through qualitative and quantitative analysis, the eight major urban clusters were further distinguished into three categories: well-developed, moderately developed, and in the early stage of development. In terms of the overall strategy of urban development, it was proposed that for the eastern and central regions, development should be moderately concentrated and balanced, and for the western region, concentrated development with clear priorities should be the norm. In addition, the hierarchical structure and functional structure of the cities were analyzed.

The study has played a positive role in promoting the analysis of the urban spatial organization at the macro level and provided some useful experience on how to grasp the spatial development at the macro scale. In particular, the analysis of the eight urban clusters (although some are far-fetched) contributed to the close connection between industry, transportation, and urban space. However, the study is not without limitations, that is the analysis of urban spatial structure is only conceptual, and there is a lack of analysis of the impact of the Eurasian land bridge on international economy, trade and culture, and the functional positioning of the central cities wants an international perspective.

(3) Cross-Century Research of China's Urban Development

A key scientific research project of the former Ministry of Construction in the *Eighth Five-Year Plan* period, the "Cross-Century Research on China's Urban Development", published by the Commercial Press in 1999 under the title *Economic Globalization and China's Urban Development*, serves as a preliminary study for a new round of national urban system planning. The book makes a comprehensive analysis of the trend of urban development in the world after World War II, studies urban development in the context of globalization and informatization and puts forward targeted measures for China's urban development.

The research first analyzes the changes in urban development model of the western developed capitalist countries since the 1950s. First, the development is trending towards metropolitan area, that is, with the progress of industry and science and technology, the population, capital, and technology are concentrated in large cities and their surrounding areas at a relatively fast speed. Meanwhile, with the improvement of urban traffic conditions, cities have gradually moved to suburban areas instead of gathering in the central region. This two-way movement accelerated the rapid expansion of large cities, resulting in a huge regional complex with distinct spaces of different levels, regional division of labor and landscape characteristics. Second, the development of metropolitan contiguous area. In economically developed areas, the relatively independent large cities gradually evolved into large-scale metropolitan contiguous areas, examples including Boston-Washington, Chicago-Pittsburgh, San Francisco-Los Angeles in the United States, London-Birmingham-Manchester in the

United Kingdom, Paris-Lyon in France, Randstad in the Netherlands, etc. These areas are believed to have played an irreplaceably important role in the economic development of countries and regions. But it is also because of the strong pull of these regions that a host of problems cropped up including the destruction of regional ecological balance, the imbalance of national and regional economic development, and the decline of central cities. So, some developed countries (such as Japan and the Netherlands) resorted to national spatial planning to redress the imbalance since the 1960s.

The research expounded in detail on the role of information society and economic globalization in urban development, upholding that with the advent of information society, the following new trends in world urbanization will emerge: first, the formation of the system of global cities, that is a number of pivotal informatized global cities developed into world cities, dominating the world's economy; second, the metropolitan contiguous areas showed more dynamics for development. From a macro point of view, the geographical pattern of "concentration in major areas while minor dispersion exist in each major area" of cities will exist for a long time to come. As a result, human settlements worldwide will gradually migrate to regions with good climatic conditions and quality of living environment. A multi-polar and multi-level network system of world cities formed, and the development of cities is increasingly dependent on the intensity of its interaction with other cities around the world and the degree of synergy they managed to create. These analyses are important for us to gain a deeper understanding of the development trend of urban built-up areas during the rapid urbanization in China and serve as important reference for us in our efforts to cultivate cities and urban areas with international competitiveness in the era of globalization.

Against the macro background of economic globalization, the research predicts the trend and prospect of China's urban development: more new cities will emerge as a result of the urbanization process of unprecedented level; large cities will continue to develop rapidly, and several metropolitan contiguous areas are already in the making; common prosperity realized in both the suburban areas and the downtown areas of cities; several international cities will become part of the network system of global cities. In view of these trends, the research recommended the following policies: (1) Building the four cities of Shanghai, Beijing, Guangzhou, and Dalian into world-class cities to be integrated into the world city system; (2) Attach importance to the construction of metropolitan contiguous areas, focusing on the four areas in the Yangtze River Delta, the Pearl River Delta, Beijing-Tianjin-Tangshan, and Central and Southern Liaoning, and stress the coordinating role of regional planning; (3) Construct a two-tier local government mechanism in metropolitan areas and implement the model of establishing cities under counties; (4) Encourage and promote the process of urbanization, and reach the urbanization level of moderately developed countries by 2020, that is, about 60%; (5) Look at the phenomenon of suburbanization in an objective and sensible way and gradually improve the policy of suburban development; (6) Accelerate the construction of inter-city fast transit network; (7) Strengthen the macro-control of urban system planning, establish the concept of

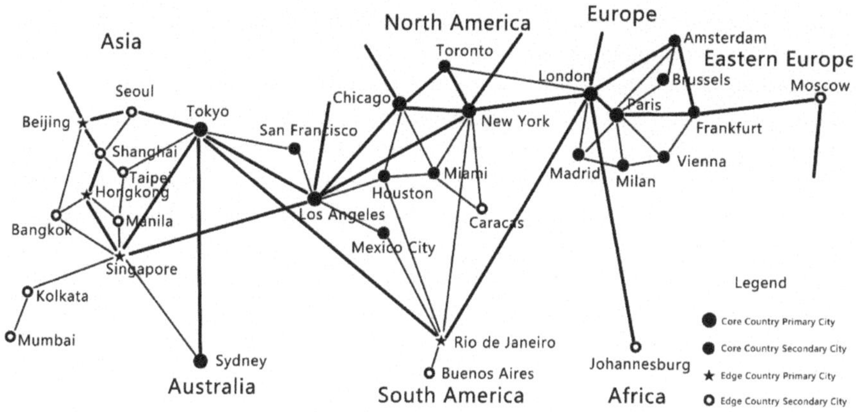

Fig. 4.12 Global urban system. *Source Economic globalization and world urban development*, [10]

master planning, and enhance the government's capability to intervene; (8) Formulate urbanization development strategies by zonings; (9) Attach importance to urban development with limited resources, and properly handle the relationship between rural industrialization and urbanization; (10) Build a national urban system compatible with the national strength, and determine the status and development goals of cities at both the global and local levels.

The research and the ten major development policy recommendations it proposed basically encapsulate the general development ideas of China's urbanization in the era of globalization, which is more contemporary than any previous research on China's urbanization. Its analysis of the spatial structure of Chinese cities and towns in the context of the world's urban system is particularly pioneering. However, the limitations of the research lie in the fact that the analysis of the world situation from the perspective of China is insufficient, or there lacks analysis on how integrate the global trend of development with the actual situation of China, and in terms of spatial structure, the research on the construction of the urban network system and its supporting system and on the constraints of environment on development are basically conceptual. Objectively speaking, it is a valuable study (Fig. 4.12).

4.3.3 National Urban System Planning in 1999

As part of the work of the former Ministry of Construction, the preparation of the national urban system plan was officially launched in 1999, and the result was a relatively complete report after almost five years of hard work. The report put forward a comprehensive plan for national urbanization and urban development from seven aspects: project background, status of urbanization and urban development, strategic objectives of urban development, planning of urban spatial distribution, urban development and transportation, coordination of resources and environment, and guarantees of implementation.

According to the plan, China is in a transition period of comprehensive social and economic development, the implementation of a positive urbanization strategy is the guarantee for realizing modernization in the long run, and the key to fixing the social problems at present and in the near and medium future, such as promoting the development of the rural economy and the tertiary industry, growing investment demand, and preserving resources. The guiding principles proposed by the planning are: analyze the laws of world economy development and global urbanization from a global perspective; consider population migration and urban spatial distribution during urbanization from the perspectives of sustainable development and preserving important ecologically sensitive areas; attach importance to regional coordination and actively implement the national strategy of "development of the western region", focusing on promoting the development of cities in the central and western regions; stress the leading role of cities and towns in social and economic development, and gradually establish a spatially well-organized urban development network to support the development of the national economy; and maintain balance between the planning of national economic and social development and urban development, between industrial layout and urban spatial organization, and between regional infrastructure and urban spatial organization. The guiding principles of the plan are deeply influenced by the central government's policy of "large-scale development of the western region" at the time. Although it proposed objectives from a global perspective, its contents were mainly focused on the development of the western region.

The priorities of the plan are to: assist the government in adapting to the transformation of functions, and provide the basis and means of macroeconomic regulation and control for the central government to guide the overall coordinated development of cities and towns across the country; effectively implement the national urbanization development strategy through national urban system plan, and promote coordinated development of the economy and society through the rational guidance of urban spatial distribution, focusing on coordinating cross-regional development issues; on the premise of fairness, share regional infrastructures through reasonable and proper organization, reduce the cost of regional development, and put forward requirements concerning regional infrastructure layout such as large airports, ports, highways, and railways, and spatial requirements regarding land use, ecological environmental protection, and water resources development. Although the measures of resource management are more theoretical, the focus of the plan is on the guidance of policy and the establishment of coordination mechanisms.

In determining urban spatial development policy, given the obvious regional differences in natural conditions and level of economic development, resource shortage, and fragile ecological environment, it is proposed that China aim for diversification and differentiation of its urban development policy. In the rules of implementation, the following fresh ideas were proposed: attach importance to the development planning and construction guidance of built-up urban areas, boost the functions of large cities, and treat the development of cities and towns in the three major areas differently. In terms of enhancing city's functions, it is proposed to cultivate the three international cities of Hong Kong, Shanghai and Beijing, and support the eight regional megacities of Shenyang, Dalian, Tianjin, Wuhan, Nanjing, Guangzhou, Xi'an and Chongqing, as well as the two cities of Shenzhen and Xiamen in the special economic zone, so that they can integrate with the world economy and become the hubs of development of its region and the entire country. Support should also be given to the 17 regional mega cities of Harbin, Changchun, Shijiazhuang, Taiyuan, Baotou, Jinan, Qingdao, Changsha, Nanchang, Zhengzhou, Hangzhou, Fuzhou, Lanzhou, Urumqi, Chengdu, Guiyang, and Kunming, and the two cities of Zhuhai and Shantou in the special economic zone to turn them into innovation bases for national industrial development. In addition, large cities with a population of 500,000 to one million in general are categorized into the following four types based on their attributes: comprehensive, industrial, mining, and transportation hub.

In terms of urban spatial organization, the pattern combining points (central city), axes (urban belt) and areas (three major areas and 12 urban built-up areas) is proposed. The "points" refer to national first-level central cities and important central cities, mainly including Shenyang, Dalian and Harbin in the economic cooperation zone in Northeast China, Beijing, Tianjin, Jinan and Qingdao in the economic cooperation zone in North China, Xi'an, Lanzhou and Urumqi in the economic cooperation zone in Northwest China, Shanghai, Nanjing and Hangzhou in the economic cooperation zone in East China, Wuhan, Zhengzhou and Changsha in the economic cooperation zone in Central China, Chongqing, Chengdu and Kunming in the economic cooperation zone in Southwest, and Hong Kong, Guangzhou, Macao, Shenzhen, Xiamen in the economic cooperation zone in South China. "Axes" refer to national primary and secondary axes. The primary axes include three north–south axes: the coastline, the Beijing-Guangzhou Railway, and line composed of Baotou-Lanzhou Railway, Baoji-Chengdu Railway, and Chengdu-Kunming Railway, as well as two east–west axes: the Yangtze River and the line composed of Lanzhou-Lianyungang Railway and Lanzhou-Xinjiang Railway. The secondary axes include the following lines: Beijing-Shanghai and Shanghai-Hangzhou-Ningbo railways, Harbin-Dalia and Beijing-Shenyang railways, and Beijing-Hong Kong Railway, and the cities located on the 8 axes account for 40% of the national total. "Areas" refer to the urban built-up areas at the national level, including the following 12: the Yangtze River Delta, Pearl River Delta, Beijing-Tianjin-Tangshan, central and southern Liaoning, Shandong Peninsula, and eastern and southern Fujian in the eastern region, the Jianghan Plain, the Central Plains, central Hunan, and the Songnen Plain in the central region, and the Sichuan Basin and Guanzhong Region in the western region.

The plan also puts forward different requirements for urban development of the three major zones. In the eastern zone, in view of its superior natural conditions, ideal location, and strong comprehensive economic strength, call for the industries to relocate from large cities to small cities and towns, promote the adjustment and upgrading of the industrial structure of large cities, and give birth to several urban clusters featuring coordinated development of large, medium, and small cities. In the central zone, given it is in the transition period of development from the early stage to the middle stage of urbanization, it should aim for industrialization and modernization and take the urbanization path combining the elements of both concentrated and scattered development, developing large and medium-sized cities in a moderate way and small cities in a rational manner. In the western zone, in view of its under-developed economy, low population quality, and insufficient development of the urban system, it should aim for industrialization and focuses on transforming existing central cities and cultivating and developing new economic centers because the key to urban development in here lies in the numerous counties and industrial and mining towns in resource development areas. The plan paid special attention to small towns and held that the surplus rural labor force will pour in small towns during the planning period. It is difficult to say whether small towns will be able to shoulder so big a burden (Fig. 4.13).

Regarding the implementation of the plan, the report proposed such measures as adopting a proactive policy for urban development, promoting the development of small towns, providing guidance for urban construction in western region, conserving urban land, and reforming the household registration system.

The report covers a multitude of problems and considers all aspects of urbanization and urban development but lacks clear structure. There are no in-depth analyses of the constraints on urban development and of population migration, and there is no relevant research to support the spatial policy of giving heavy support to the western region and vigorously developing small towns, generating contradiction between development of western region and protecting the ecological environment in the

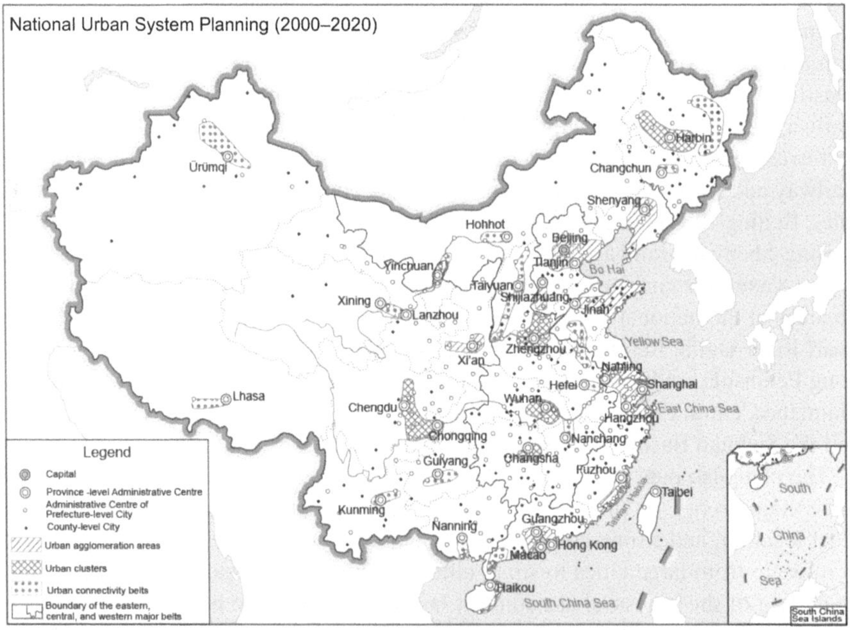

Fig. 4.13 Spatial distribution of the national urban system planning, 1999. *Source* National Urban System Planning (2000–2020), China Academy of Urban Planning and Design, 2000

western region. Many of the urban clusters proposed in the plan vary greatly in terms of level and content, and it is problematic to lump them together as national-level urban clusters. And in implementing the plan, there is a lack of concrete guidance for the lower level of planning.

4.3.4 *Outline of the National Urban System Planning 2005*

(1) Background Analysis for A New Round of Planning

As mentioned earlier, no formal document of national urban system planning was produced to be submitted to the State Council for review during the five years from 1999 to 2004. During this period, the *Outline of the Overall National Land Use Plan* compiled by the then Ministry of Land and Resources was officially implemented nationwide after being approved by the State Council, and the then State Planning Commission completed the study of "Theoretical Exploration of the Reform of Planning System" and proposed the concept of "national master plan" [11]. In October 2005, the "Proposal for Formulating the Eleventh Five-Year Plan for National Economic and Social Development" adopted by the Fifth Plenary Session of the 16th Central Committee of the CPC changed the "Five-Year Plan"("五年计划") which had been used for the past 50-odd years into "Five-Year Planning" ("五年规划") and added the content of spatial planning such as "cultivating new urban clusters". All these indicate that spatial planning has been gradually adopted as a macro-control means for national social and economic development.

It is thought-provoking why, in the absence of land resources planning and the content of the overall land use planning being too simple, a "national urban system planning", a comprehensive statutory plan, failed to be promulgated. Author of this paper believes that there are three main reasons. First, the government and the academic community do not have sufficient understanding of the role that national urban system planning can play in macroeconomic regulation and control of national social and economic development. The popular belief is that under the market economy, planning should be premised on the law of the market[17]. The country lacks a strategy for coordinated development, and the time is not ripe for planning. Although the central government adopted such regional policies as the large-scale development of the western region, the revitalization of the old industrial bases in the northeast, and the rise of the central region, there is no clear thinking on the overall coordinated development. Third, scholars in urban planning have not done sufficient theoretical preparation. In the past decade, the planning community has been focusing on how to promote rapid development of the urban economy through planning and how to plan for and design downtown areas, key areas of landscaping,

---

[17] At the National Symposium on Urban Planning in Wuxi, 1992, Zhao Shixiu proposed that urban planning would not merely be the continuation and embodiment of planning, and that cities, as the "carriers" of the economy and its various activities, would increasingly operate according to market mechanisms.

and new cities at the micro level. Spatial planning at the macro level are mainly theoretical analyses of the geographical community. In the past decade, the regional urban space research of real influence is the one on "Beijing-Tianjin-Hebei Urban and Rural Spatial Development", which proposed the spatial structure of the region from a global and regional perspective. The research has had an important impact on the study of macro-level spatial structure and contributed to the subsequent research on urban spatial development strategy and to the preparation of urban master plans in Beijing and Tianjin. No other research or planning available can match it in terms of influence and impact.

In April 2005, as part of the contents for improvement in the "Campaign to Maintain the Advanced Nature of Communist Party Members" of the former Ministry of Construction, a new round of national urban system planning was started. And in October the same year, the Party Group of the ministry reviewed and approved the planning outline. This round of planning has been attached more importance for three reasons: the "scientific outlook on development" put forward by the central government provides a new guiding ideology for urban system planning. Second, the National Development and Reform Commission's proposal for the reform of national planning system accelerated the promulgation of the planning[18]. Third, the host of problems cropped up during the rapid industrialization and urbanization is also an important reason for the fast introduction of the national urban system planning.

In September 2005, the Political Bureau of the CPC Central Committee held the 25th study meeting on the "Urbanization Development Models of Foreign Countries and the Urbanization Path with Chinese Characteristics" to discuss the development of China's urbanization. Hu Jintao, the then General Secretary, stated in his concluding remarks the requirements for China to take the path of wholesome urbanization and of speeding up the preparation of the national urban system planning. This is the first time the highest decision-maker made clear about its attitude towards national urban system planning since the promulgation and implementation of the *Urban Planning Law* in 1990.

(2) Main Technical Features[19]

Based on the trend of international political and economic development, the future industrial policy of the country, the trend of population migration and the characteristics of different regions, the planning analyzed the situation and characteristics of China's urbanization and the path of future development, attempting to guide the national industrial spatial distribution, resource preservation and regional infrastructure construction through the construction of a new national urban spatial structure. In terms of content and outcome, it attaches importance to the analysis of the prerequisites for urban development and strives to highlight the attribute of planning which is public policy. Its main technical features are:

---

[18] According to the new planning system proposed by the National Development and Reform Commission in the study "Theoretical Exploration of the Reform of Planning System", urban planning belongs to the category of special planning.

[19] Analyzed and sorted out based on the draft of the *Outline of National Urban System Planning, 2005–2020* dated October 2005.

First, strive to implement the development idea that is scientific, comprehensive, and coordinated

At the outset, the planning stated unequivocally to follow the people-oriented, comprehensive, coordinated, and sustainable scientific outlook on development, and actively and steadily implement the guiding ideology of urbanization in accordance with the principles of gradual and orderly progress, preserving land as much as possible, intensive development and rational distribution. We should give more play to the important role of urban and rural planning as a means of government regulation and control, social management, and public services, and tap into the potential of cities and towns in promoting economic and social development, to improve the quality of urban and rural living environment in an all-round way. The long-term plan (2011–2020) is consistent with the objectives of national medium—and long-term planning, and the short-term plan (2005–2010) is compatible with the national *Eleventh Five-Year Plan*. And new contents are proposed in the cultivation of spatial nodes and the management of spatial resources.

Second, the development needs of cities and towns under the new situation are fully analyzed

The planning analyzes the needs of future urban development from the perspectives of the challenges of economic globalization, implementation of new industrialization, construction of new socialist countryside, coordinated regional development, and the building of a moderately prosperous society in an all-round way. The planning holds that in the context of ever-deepening economic globalization, development will remain the norm in the next 15 years. In the new round of restructuring of global production, trade, science and technology, and culture, China should secure herself a place by adjusting spatial structure to build cities and regions that are compatible with the new reality. According to the general policy of the state to take the new road to industrialization, it is proposed that the five major urban economic zones of Beijing-Tianjin-Hebei, Yangtze River Delta, Pearl River Delta, central and southern Liaoning, and Chengdu-Chongqing, as well as ten population-industry concentrated areas such as Harbin-Daqing-Qiqihar will be built in the next 20 years.

Third, a comprehensive study of the constraints on urban development was made

The planning made an in-depth study of China's natural geographical conditions and did a preliminary analysis of the proportion of land and the main areas suitable for urban construction in China based on a comprehensive evaluation of natural conditions (see the table below). According to the analysis, the areas suitable for urban development (excluding cultivated land) account for only 8.55% of the country's land area, and the carrying capacity of the eastern region is relatively high, while that of the western region is low. This is the first comprehensive evaluation of the conditions for constructing human settlements in China (Fig. 4.14).

The planning also stated clearly that the basic cropland of 1.65 billion mu (or 330 million ha.) is the safety net of food security in accordance with the country's basic

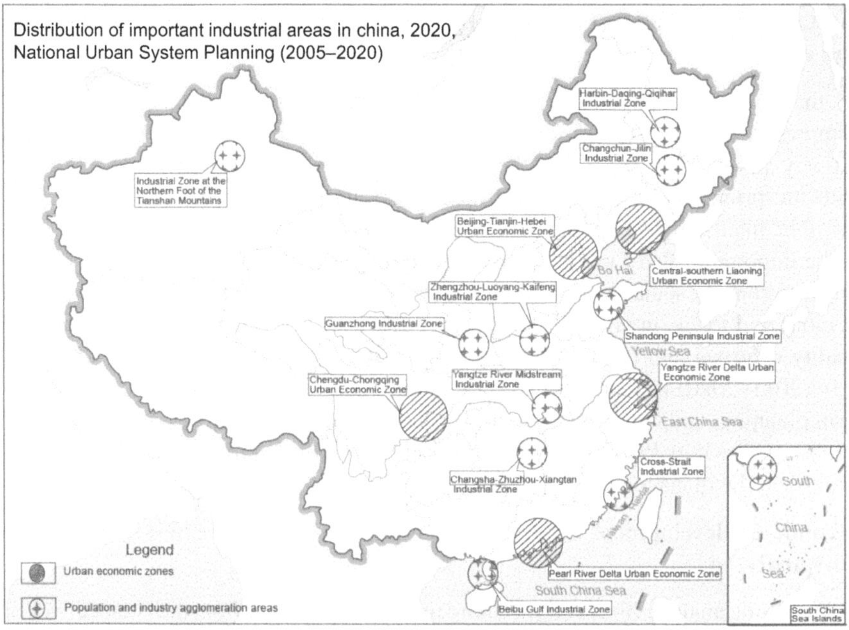

Fig. 4.14 Distribution of important industrial areas in china, 2020. *Source* Research on the Strategy of Eleventh Five-Year Plan, Ma Kai. Beijing Science and Technology Publishing Co., Ltd. 2005

national policy of protecting cropland. Facing the reality that the areas suitable for urban construction are often the main grain-producing areas, specific requirements were put forward to handle well the relationship between urban and rural construction and cropland protection. The planning also made a thorough analysis of population growth and its flow trends, proposing that the scale of population flow and migration will increase in the future. The coastal areas will remain the first destination of the floating population before 2015, while the central region, with economic development, will change from an area of net population outflow to an area with a balance of population inflow and outflow around 2015 (Table 4.3).

Fourth, differentiated urbanization policies of the eastern, central, and western regions were determined

Based on the different development requirements of the state for the eastern, central, and western regions, as well as the unique resources and environmental conditions of these regions, the planning put forward policy guidelines of urbanization for each region for the first time. To be specific, the planning proposed separate spatial strategies for the eastern, central, western, and northeastern regions from the perspectives of overall urbanization requirements, priorities in spatial development, infrastructure supply and environmental protection measures. For instance, the development guidelines for the eastern region are to improve the quality of urbanization and optimize the demographic structure; accelerate the development and resource integration of the

Table 4.3 Summary of the suitability of land areas for urban development

Suitability	Areas	Account for (%)	Proportion of cropland (%)
Entirely unsuitable	Mainly in the western region, Inner Mongolia in the central region, the southwest region, and the tidal flats and wetlands in the east, including the Taklamakan Desert, the Northern Xizangan Plateau, and the central Gobi in Inner Mongolia. These areas are mainly restricted by such factors as elevation, slope, soil erosion, and precipitation	52	5
Relatively unsuitable	Mainly in mountainous and hilly areas with steep slopes, semi-arid areas with little precipitation, areas with low accumulated temperature that are not suitable for crop growth, and areas largely covered with forest and grass	29	20
Suitable	Flat in terrain and rich in water resources, they are mainly in plains and basin areas and used largely as cropland. To be specific, they include the Northeast Plain, Sanjiang Plain, North China Plain, the middle and lower reaches of the Yangtze River Plain in the eastern region, as well as the Sichuan Basin, and the Hexi Corridor in the west, and the alluvial areas in the north and south sides of the Tianshan Mountains. It is the main space for the future urban development of China	19	55

Source Outline of National Urban System Planning, 2005–2020

three major metropolitan contiguous areas and curb the sprawl of space; strengthen the protection of ecological environment, especially the comprehensive management of water environment, and build water-saving cities.

In the prediction of urbanization level, the planning, based on the average rate of urbanization in China in the past 20 years and the trend of social and economic development in the next 20 years, indicated that urbanization will plateau at an average annual growth rate of 0.8 to 1 percentage points in the next 15 years, and the figure will further reach 46 to 48% in 2010 and 55 to 58% in 2020. In addition, the planning also gave a forecast of the urbanization level of each province based on the economic level and population growth trend of each province (Fig. 4.15).

Fifth, the multi-center spatial structure is proposed

The urban spatial development strategy is the core of the national urban system planning. The planning proposed a multi-center spatial structure to promote the coordinated development of different regions in a comprehensive way. Specifically, the structure comprises one belt, seven axes, and multiple centers. "One

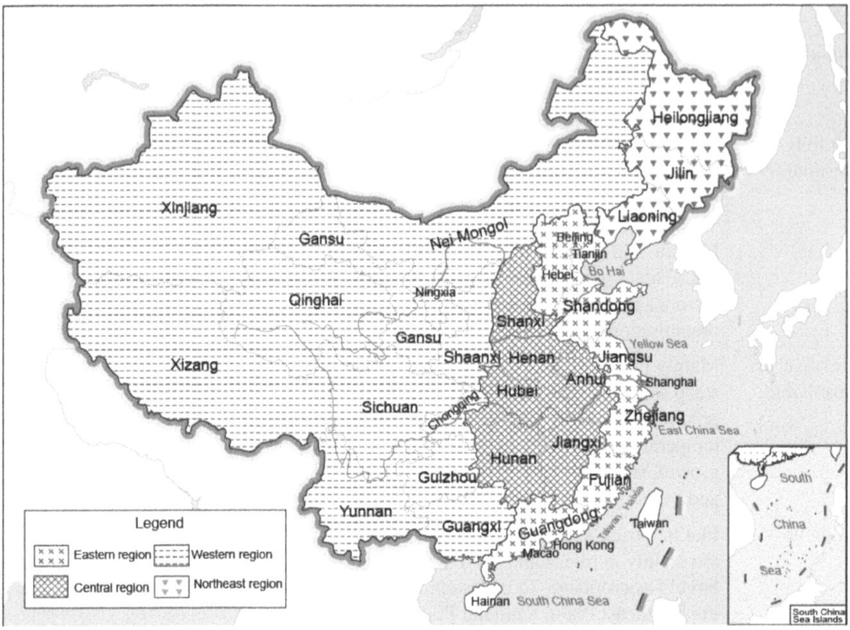

Fig. 4.15 Policy zoning diagram of urbanization

belt" refers to the metropolitan belt along the coast. "Seven axes" refer to the four east–west axes of Beijing-Hohhot and Baotou-Yinchuan railways, Lianyungang-Lanzhou and Lanzhou-Xinjiang railways, the axis along the Yangtze River, and Shanghai-Ruili, and the three north–south axes of Beijing-Guangzhou Railway (part of Beijing-Kowloon Railway), Harbin-Dalian Railway, and Beibu Gulf. "Multiple centers" refer to the major metropolitan contiguous areas of Beijing-Tianjin-Hebei, the Yangtze River Delta, and the Pearl River Delta, as well as the urban clusters of Wuhan, Chengdu-Chongqing, central Liaoning, Guanzhong area, Shandong Peninsula, Zhengzhou, Changsha-Zhuzhou-Xiangtan area, and the west bank of the Taiwan Strait. Among them, Wuhan, Guanzhong area, central Liaoning and Chengdu-Chongqing are the key urban clusters that would be used to boost the economy of Central, Northwest, Northeast and Southwest China. The concept of "urban cluster" is a new approach adopted by the planning to break through the previous urban spatial structure based on individual cities.

The planning did not dwell on the hierarchical structure of cities, but stressed the cities that need to be given priority for the interest of the overall national development, such as land gateway cities, old industrial base cities, mining (natural resources) cities, old revolutionary base areas, cities and towns in ethnic minority areas, and national historical and cultural cities and towns. Considering the requirements of building new socialist countryside, the planning put forward the following principles of village and town spatial distribution: guide the rural population to move to key towns and central villages with better conditions and improve the service of infrastructure and public facilities in those towns and villages (Fig. 4.16).

Sixth, the concept of integrated transportation hub was proposed

The support system is crucial for urban development, so, the planning proposed to build a support system with transportation as the core. Considering the spatial structure comprising "one belt, seven axes and multiple centers" and six major transportation zones, seven groups of first-level integrated transportation hub cities were established to promote an integrated combined transportation mode. The status and transportation service level of gateway cities were improved to promote the organic integration of intra-city and inter-regional transportation. In addition, for the special transportation plans of civil aviation, railways, highways, and ports that have been completed by now, adjustments might be made from the perspective of regional and urban development to ensure compatibility with new urban spatial structure (Fig. 4.17).

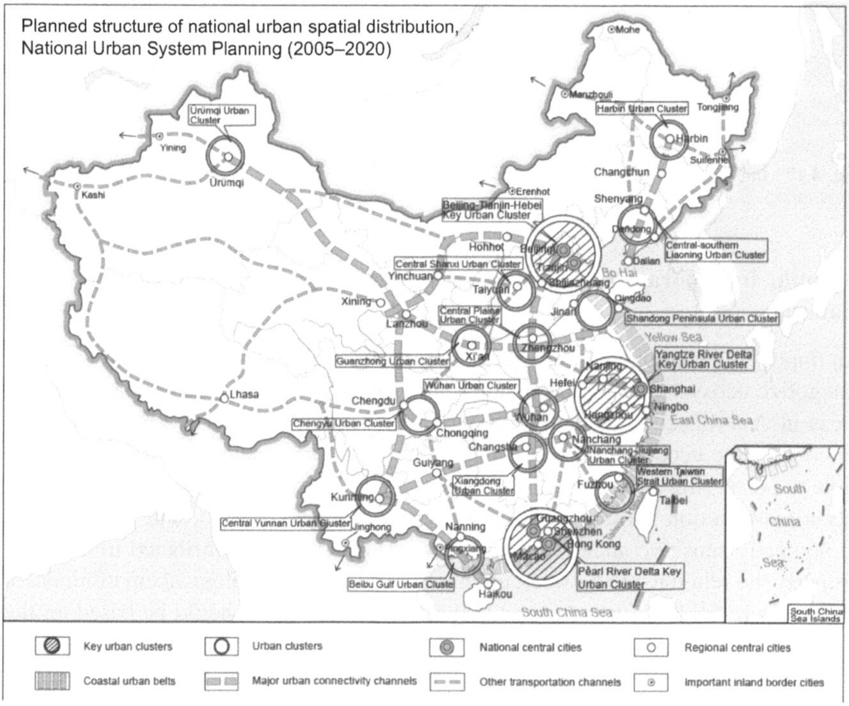

Fig. 4.16 Planned structure of national urban spatial distribution. *Source Outline of National Urban System Planning, 2005–2020*

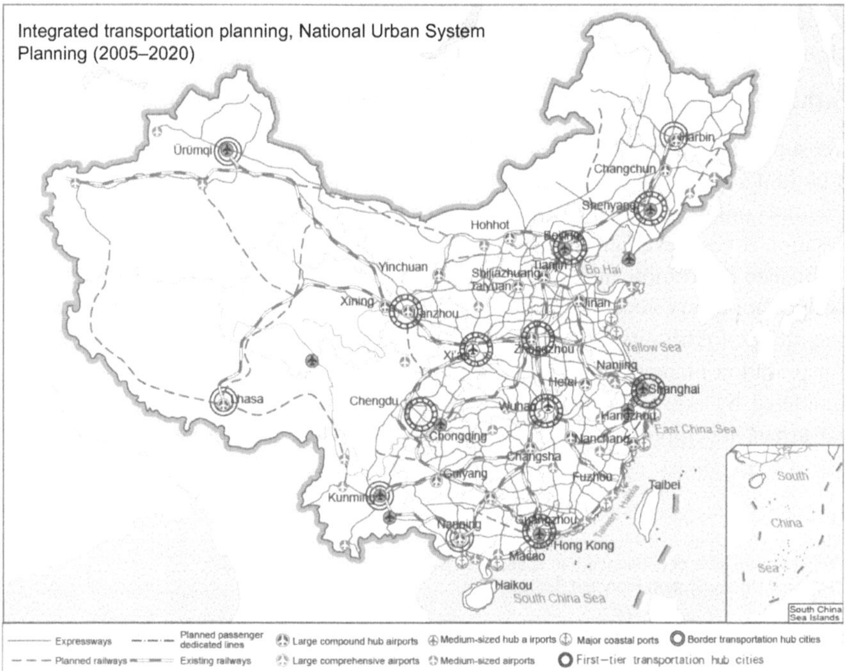

Fig. 4.17 Integrated transportation planning. *Source Outline of National Urban System Planning, 2005–2020*

Seventh, incorporate the concept of ecological security into macro spatial planning

An important idea upheld in the planning is to break the administrative division and categorize ecosystem into conservation areas, restoration areas and reconstruction areas in accordance with the requirements of ecological security and the characteristics of the ecosystems across the country. In addition, to implement the principle of preserving ecological environment, specific requirements in the selection of urban construction sites, prevention and control of soil erosion, and protection of ecologically sensitive areas were put forward, which have contributed to handling properly the relationship between urban construction and ecological environment at the macro level. It is further stated that urban construction should be based on the carrying capacities of the environment (atmospheric environment and the maximum pollutant treatment capacity of waters), and fully consider the possible impact of urban development on the ecological environment.

Eighth, separate development guidelines for provinces and autonomous regions are proposed based on the division of powers

The planning expounds on specific guidance for regional development. Two types of spatial guidelines are proposed based on the overall requirements of national urbanization and spatial distribution: first, for the cities and regions earmarked for key development and management, i.e. the regions and cities covered by "one belt, seven axes and multiple centers"; and second, the areas for key cross-provincial coordination, which mainly includes the following types of areas: river basins, coastal and offshore areas, border areas of provinces and regions, metropolitan contiguous areas, urban clusters, and mineral and other resource utilization areas.

On this basis, the planning guidelines for 27 provinces and autonomous regions and 4 municipalities directly under the central government were put forward. Take Hebei Province for example, the cities earmarked for key development and management are Shijiazhuang, Tangshan, Handan, Baoding, and Zhangjiakou, and the areas for cross-provincial coordination are the water conservation area in the upper reaches of the rivers in Beijing and Tianjin and the ecological environment of the Bohai Rim. In addition, the Three-North Shelter Forest was built; the cooperation between Qinhuangdao Port, Tangshan Port, Huanghua Port and the ports of Tianjin, Liaoning and other provinces and cities were strengthened; the high-speed passenger lines including Beijing-Guangzhou and Beijing-Shanghai railways were reserved; and the construction of Beijing-Tianjin-Tangshan intercity rail transit was accelerated. Proposing separate development guidelines for different provinces and regions is an important means to implement the national urban system planning, and putting forward control requirements by province is in line with the principle of division of powers, which contributed greatly to the implementation of the planning and to the cross-provincial preservation and management of resources.

In summary, the new round planning of national urban system made a useful exploration of macro-level spatial planning in terms of theory and methodology. It is worth mentioning that innovation was made in the analysis of the prerequisites for urban development, the diversified policy zoning of urbanization, the urban spatial structure of multi-center and resource protection, and the separate development guidelines for different provinces and region. However, due to the limited time for preparation and insufficient understanding of the planning, there is still a lot of room for improvement in theory and methodology. For the technology in general, this author believes that there are five deficiencies:

First, from the perspective of international experience, any national spatial planning should have problems it aimed at fixing, set priorities for the regional policy at the time, and achieve the development goal of "polarization" or "balance" through new spatial structure. The goals of this national urban system planning, however, are not clear in terms of regional policy and spatial organization.

Second, in a period when sustainable development has become the norm around the world, the focus of spatial planning is largely on the protection and effective use of space resources. In this regard, the planning lacks analyses of the hierarchy and efficacy of spatial resources and of environmental carrying capacity and corresponding measures in the face of rapid development, and there are no enforceable measures for the spatial development in ecologically fragile areas.

Third, any urban spatial structure is only valid for a certain period, and there is no such thing as an all-time-applicable structure. Given today's economic globalization, we need an open and flexible spatial structure. Although the planning has a relatively open one, it wants in adaptability. There also lacked guidance in terms of spatial policy for some areas that require special attention at the national level (e.g., urban built-up areas like the Yangtze River Delta).

Fourth, the central government's planning should embody the national interest and equitable services for different regions. The establishment of a unified, fair, comprehensive, and urban–rural integrated public service system should be an important part of national spatial planning, which is also an important means to prevent the marginalization of backward areas. But the planning lacked substantial content and standards of the construction of public service system, especially, it failed to pay due attention to public services in the vast rural areas.

Fifth, urban space is a comprehensive spatial system, and the planning of urban system comprises guidance of various special plans. The current planning lacks connection with the "medium and long-term planning for national economic and social development" formulated by the National Development and Reform Commission and the "national overall land use planning" organized by the then Ministry of Land and Resources, so it is not a comprehensive spatial planning in the real sense.

4.4 Summary

From a historical point of view, the development of urban space is directly affected by the political and economic system, and its birth and development are inseparable from the political and economic situation of the time and space, as well as the natural, geographical, environmental, and other conditions. In the process of industrialization and urbanization, correct guiding ideas and rational means of planning can contribute greatly to the country's economic and social development; without them, the country would suffer heavy losses. The success of the 156 key projects and the outcomes of the "Third Front" construction provided us with valuable experience and profound lessons.

Urban spatial planning plays an irreplaceably important role in the reconstruction of urban spatial structure. Over four decades since the Reform and Opening Up, China owed its sustained and rapid development largely to the national spatial development strategy of opening the coastal areas and centering on cities. In particular, the industrial spatial layout with the development zones as the mainstay that was closely integrated with the development of coastal central cities had effectively contributed to the development of regional economies.

The national urban system planning, within the current planning system, is a relatively complete national spatial development strategy and enjoys legal status. But strictly speaking, it is not a comprehensive and authoritative national spatial planning. Under the socialist market economy, the comprehensive and coordinated development of the social economy of our country is inseparable from the state's macroeconomic regulation and control, and national spatial planning, as an important means, will only become more and more important with the gradual perfection of our economic system. Now that China is integrating into the world in all aspects, it is worth exploring, both theoretically and practically, how to give full play to the role of urban and regional planning in social, economic, and regional development through the approach of spatial planning.

References

1. Wu LY (2003) Zhang Jian and Nantong as the first modern city of China. City Plann Review 7:6–11
2. Cheng YF, Cheng YH (1984) Preliminary compendium of historical archives by resources committee (Volume I). Taiwan: "National Museum of History"
3. Dong ZK, Wu J (2004) Industrial cornerstone of new China: a study of 156 construction projects. Guangdong Economy Publishing House 2
4. Chen DS et al (1993) Regional economics. Henan People's Publishing House
5. Lu DD (1990) Theory and practice of industrial layout in China. Science Press, Beijing, pp 22–27
6. Bi WM (1994) Land remediation and economic development. Capital Normal University Publishing House, Beijing
7. Wu CJ, Hou F (1990) Land resource development, remediation and planning. Jiangsu Education Publishing House, Nanjing, pp 378–379
8. Cui GH et al (2002) Regional analysis and planning. Higher Education Press, Beijing, 1, pp 345–346, 353
9. Wang J (2002) Comparison and reference of land planning systems in Japan and ROK. China Land Sci (3):45–48
10. Gu CL (2000) Economic globalization and urban development in China. The Commercial Press, Beijing, pp 56–60, 319
11. Yang WM (2003) Theoretical exploration of planning system reform. China Market Press, Beijing, p 6

Chapter 5
Theory and Methods for Urban Spatial Development in China

5.1 Macro Background of Urban Development in China

After more than 20 years of rocketing economic growth and rapid urbanization, China has reached a turning point in development. With per capita GDP reaching $1000 by the end of 2004, China has entered a stage of major social transformation which has been recognized by the international community. The level of urbanization has reached 41.7%, entering the period of rising urbanization in the "S"-shaped curve portrayed by R. Northam, an American geographer. Although there are different views on the path of China's urbanization and future development, there is a broad consensus that China's urbanization will continue to grow fast in the foreseeable future, or 15–20 years to come[1]. J. Stiglitz, the 2001 Nobel Laureate in economics, even juxtaposed the urbanization in China and the high-tech industry in the U.S. as the two major events affecting the world economy in the twenty-first century.

However, China's choice of the future path to urbanization should be based on the objective judgment of the historical stage of national development and the objective analysis of international environment and constraints to development. Firstly, China's productivity is still not highly developed, and it will remain in the primary stage of socialism for a long time, with development as the main theme. Secondly, international competition is becoming increasingly fierce, and continuous improvement of international competitiveness is the need of the times. Thirdly, the pressure of resource scarcity and environment in the future urbanization process is heightening, highlighting the imbalance between urban–rural development and across regions. For development, it is imperative to set the overall strategy of China's urbanization from the perspective of proper utilization of resources. Therefore, to coordinate population distribution, industrial development, infrastructure construction, environmental protection and balanced regional development at the spatial level is fundamental to the goal of creating a sustainable living environment.

---

[1] The *11*th *Five-Year Plan* for national economic and social development, promulgated in March 2006, estimates that the level of urbanization in China will reach 47% by 2010.

K. Wang, *National Spatial Planning in China*, Urban Sustainability,
https://doi.org/10.1007/978-981-97-7729-7_5

5.1.1 Impact of Economic Globalization on Urban Development

As mentioned earlier, economic globalization will deepen in the next 20 years when China is going to face an important strategic opportunity to accelerate the pace and improve the quality of development. In the context of globalization, the new round of global shift enables China to take a place during the reconstruction of global production, trade, technology, culture, and investment network. It suffices to say that building cities and urban areas with international competitiveness is an important means and measure for China to improve international competitiveness and comprehensive national power. At the same time, in the context of globalization, it is inevitable for China to strengthen the links among domestic industries and coordinate the specialization of labor among regions to integrate the regions into the global economic system as soon as possible.

At present, the major international cities in the world, such as London, New York, Tokyo, and Paris, constitute the core of the global city system, playing the function of organizing, controlling, and managing the regional economy and even the world economy. In recent years, as international institutions and multinational organizations continue to cluster in Beijing, Tianjin, Shanghai and Guangzhou, these cities are gradually taking the shape of international cities. The mega city regions formed around them, such as Beijing-Tianjin-Hebei, Yangtze River Delta, and Pearl River Delta, have gradually become the mainstay of China's participation in international competition. The development of these core cities and their surrounding areas in the next 20 years will largely determine the international competitiveness of China and its position in the world economic and political landscape.

At the same time, it should be noted that the trend of regional economic integration is prominent. In recent years, China has further strengthened cooperation with neighboring regions and become a member of several international regional cooperation organizations such as Asia–Pacific Economic Cooperation, Shanghai Cooperation Organization, Southeast Asian Economic Association, and Greater Mekong Subregion. The continuous development of bilateral and multilateral trade highlights the importance of developing border cities and international transportation corridors. Border crossings in the northeast, northwest and southwest are gradually evolving from mere border trade centers to large-scale and intensive international entrepot trade centers, international transportation hubs and export-oriented manufacturing centers. With the establishment of China-ASEAN Free Trade Area, Central Asia Free Trade Area, and China-Japan-Korea Free Trade Area in the future, there is no doubt that the development of cities in the coastal and border areas will reach a new height. Only the cities and regions that "embrace the two markets and use the two resources" can really achieve great development in the international competition. The interdependence of international market and domestic market requires us to strengthen the connections between coastal and inland areas in terms of economy, trade, industry, and transportation, cultivate the inland market and fully integrate into the world economy.

5.1.2 Impact of National Industrial Policy and Industrial Landscape on Cities and Towns

To take the path to new industrialization and develop agglomeration economy, circular economy and knowledge economy is a long-standing policy that China will adhere to in the twenty-first century [1, 2]. At present, the general trend of China's industrial development is the transformation from general processing industry and light industry to intensive and specialized processing industry and heavy chemical industry, and the industries tend to centralize their organization. Studies show that in the next 20 years, China will form five major metropolitan economic zones, including Beijing-Tianjin-Hebei, Yangtze River Delta, Pearl River Delta, South Central Liaoning, and Chengdu-Chongqing, as well as ten population-industrial agglomerations, including Harbin-Daqing-Qiqihar, Changchun-Jilin, Shandong Peninsula, Both Sides of Taiwan Strait, Zhengzhou-Luoyang-Kaifeng, Central Sha'anxi, Middle Reach of Yangtze River, Changsha-Zhuzhou-Xiangtan, Beibu Gulf, and Northern Slope of Tien Shan [3]. In general, the heavy chemical industry such as petrochemical, iron and steel will shift to consumption hubs and coastal and riverside cities that are in a better position to use imported resources; and the energy industry, especially the electric power industry, will gradually shift to western cities. The eastern region will focus on the development of high-tech and capital—and technology-intensive industries, while the central and western regions will pivot around traditional industries and labor-intensive industries, which will be the division of labor across the regions for an extended period.

At the same time, with the increasing demand for independent innovation in science and technology, strengthening the development of technology-intensive regional central cities is critical for improving China's comprehensive national power and upgrading the industrial structure. Promoting the agglomeration of enterprises and fostering the growth of innovative cities and regions is an important foundation for local economic development. As the process of industrialization continues, the tertiary industry is taking up a higher proportion in the economic structure and becoming increasingly important. The modern service industry will be rooted in central cities. The high-end financial, logistics and R&D functions from across the country will be further concentrated in Beijing, Shanghai and Guangzhou who will become the organizer and manager for national economic growth with stronger influence and service capacity.

5.1.3 Impact of Population Migration on Urban Development

According to the Fifth National Population Census in 2000, the size of population migration across China was 144.39 million, accounting 11.6% of the national total. Given the general principle that when the internal migrants stabilize at 10% or more of the population, the country or region is in a period of migration, it can be concluded that China is entering a period of migration. The arrival of the migration period means a huge number of people moving from less developed regions to developed regions, from rural areas to cities and shifting from inefficient production to highly efficient wealth creation, which reflects the great economic and social changes and development in China as well as an epitome of the massive urbanization in the country (Fig. 5.1).

A study shows that in the next 20 years, population movement in China will continue to follow the basic pattern of migrating from rural areas to cities, from less developed regions to developed regions, and from small and medium-sized cities to larger cities. With the reform of household registration system, the proportion of long-distance migration will tend to rise, reaching 35%; the proportion of short-distance or inter-district and inter-county migration will drop greatly, but will remain at about 30%; and the proportion of medium-range migration will increase to 35%. The coastal areas will remain the first choice for internal migrants up to 2015, and then the proportion of the floating population absorbed by the coastal areas will remain stable, but the proportion will drop slightly from 2015 to 2020. Around 2015, the inflows and outflows in the central region will reach an equilibrium due to economic development. metropolitan interlocking region and urban clusters will be the main space to absorb the people [4].

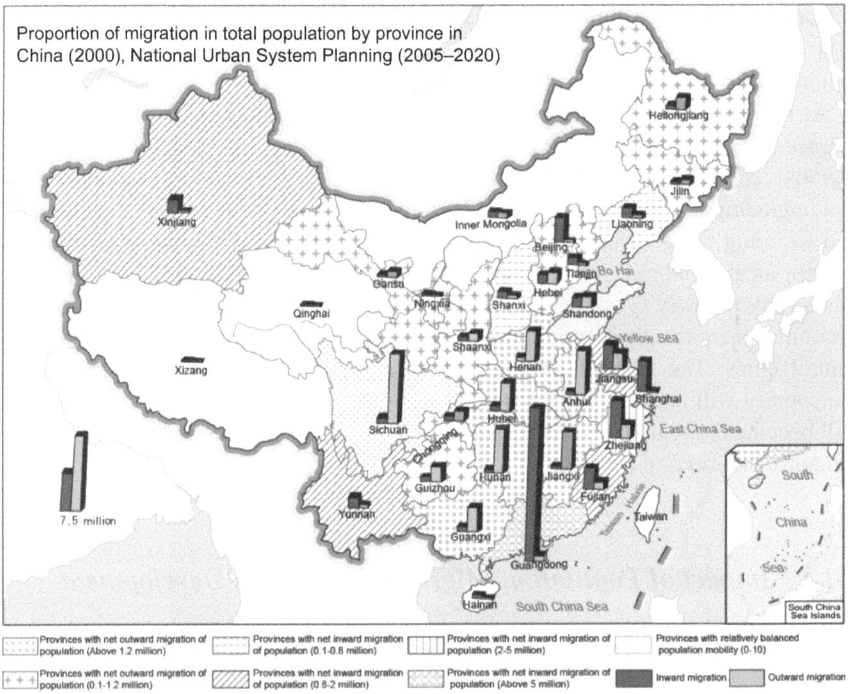

Fig. 5.1 Proportion of migration in total population by province in china. *Source* Outline of National Urban System Planning, 2005–2020

According to the projections of the former National Family Planning Commission, China's population will reach a peak of 1.5 billion around 2033 based on current total fertility rate of 1.8. Before 2033, the population increase is estimated at 8 to 10 million people per year, and what's significant is that the working-age population, the total population, and the population aged 65 and over will hit their peaks[2]. The structural demographic problems will become prominent, the scale of population movement and migration will increase, and the contradictions between population, resources and environment will become increasingly acute. Considering the process of urbanization in the past 20 years and the trend of future development, the level of urbanization in China will further increase, and experts predict that the level of urbanization will reach about 55% by 2020 [5].

5.1.4 Resource and Environmental Constraints on Urban Development

Objectively speaking, the development of cities is constrained by geographical conditions, environmental carrying capacity, land, and water resources. The geographical conditions cover such aspects as topography, climate, temperature, and land type. The areas unsuitable for human survival and settlement account for 52% of the national territory, mainly in the western region, i.e., west of the "Aihui-Tengchong line"; the less suitable areas account for 29%; and the suitable areas account for 19%, mostly in the eastern and central regions. The areas suitable for human settlement have the highest concentrations of arable land in China. The spatial distribution of cities is constrained directly by geographical conditions (Table 5.1).

In terms of water resources per capita, China ranks 121st in the world and is generally considered as a water-scarce country. The spatial distribution of water resources is extremely uneven. Water resources in the Yangtze River basin and the areas to the south account for 81% of the national total, whereas the Yellow River, Huai River, and Hai River basins in the north only account for 7.2%. The surface water in the Hai River and Yellow River basins is already in a state of over-exploitation, and the available surface water resources in the Huai River basin and the rivers in the northwest are being stretched to the limit; and the ground water in Hai River basin, Huai River basin and Shandong Peninsula has been excessively withdrawn. In terms of urban water use, the cities in the eastern and central regions account for 87% of the national total. With the development of specialized processing industry, the amount of water needed for industrial production will continue to grow soon; and with the improvement of living standards in cities, the water demands of urban residents will also increase. According to studies, the total urban water consumption is expected to increase by 90% by 2030, compared to the 2003 level. The distribution of water resources affects the scale and mode of urban development. Although water

[2] According to the "Special Report on Population", Outline of the National Urban System Planning, 2005–2020.

Table 5.1 Indicators for human settlement evaluation

Evaluation factor	Entirely unsuitable	Relatively unsuitable	Suitable
Elevation	> 4000 m	2000–4000 m	< 2000 m
Average annual precipitation	< 50 mm	< 200 mm	> 200 mm
≥ 10 °C Accumulated temperature	< 1300	1300–1600	> 1600
Types of land use	Sandy land, Gobi, saline-alkali soil, marshland, glacier and permanent snow, water body, mudflat	Wooded land, high cover grassland	Others
Soil erosion	Extremely strong wind erosion, strong freeze–thaw erosion	Moderate freeze–thaw erosion, strong wind erosion	Others
Terrain (slope)	> 25 degrees	15–25 degrees	< 15 degrees
Topography	Extremely rolling mountain, greatly rolling mountain, sand dune, snowy plateau	Moderately rolling mountain, karst mountain, loess ridge, high-elevation hill, medium-elevation hill, low-elevation hill, karst hill, high terrace, medium terrace, micro highland	Low terrace, undulating plain, sloping plain, flat plain, micro depression

transfer projects can solve part of the problem, the numerous problems related to environmental conservation and cultural relics protection issues cannot be stamped out.

As for land resources, according to the statistics of the Ministry of Housing and Urban–Rural Development, the area of urban and rural construction land in China was 204,000 km2 in 2004, accounting for 2.1% of the country's total land area, and the per capita construction land was 158.4 m2. Compared to other countries, the level of intensive urban development in China is quite high. However, in recent years, reckless construction of development zones in some places have caused a considerable waste of land resources. By the end of July 2004, there were 38,600 km2 of construction land planned for development zones at all levels in China, equivalent to the sum of the construction land in the cities and counties nationwide.

Meanwhile, it is a basic national policy to protect basic cropland in China. At present, the total amount of arable land in China is 1.2 million km2, and according to the requirement, 110 million km2of basic cropland is the bottom line for food security. The current use of arable land is close to the bottom line. To guarantee the quantity of basic cropland, the major grain commodities production bases in the

Huang-Huai-Hai region, Songnen Plain, Sanjiang Plain, Central Jilin Plain, Hetao Plain, Ningxia Plain and Hexi Corridot will play an increasingly prominent role in increasing crop yields; and the Poyang Lake Plain of Jiangxi, Dongting Lake Plain of Hunan and Chengdu Plain of Sichuan will remain the important food producing areas in China. In the future, the regions with low crop yields and food scarcity, including Dingxi of Gansu on the Loess Plateau, Xiji, Haiyuan and Guyuan of Ningxia, Ankang and Shangluo of Sha'anxi, the west of Hubei, remote areas of Sichuan and remote mountainous areas of Yunnan and Guizhou, will be given priority in the development of agricultural infrastructure. In fact, many of the above regions play an important role in China's future urbanization.

In terms of the carrying capacity of the ecosystem, although there is no comprehensive analysis, the current levels of environmental pollution demonstrate the serious challenges facing the development of cities. The pollution of water environment has spread from land to coastal waters. In particular, the coastal waters near the three densely populated urban clusters—the Beijing-Tianjin-Hebei region, Yangtze River Delta, and Pearl River Delta—are suffering from the worst coastal water pollution and red tide in China, which requires the coastal areas to change the mode of development.

As China suffers from severe air pollution, cities and regions are often blanketed in haze, especially the North China, Central Plains, South China, and East China, where air pollution is obviously regional. Haze is more frequent in the urban clusters. Acid rain mainly takes place in Central China, Southwest China, East China, and South China, and is worst in Central China, in particular Hunan Province and Jiangxi Province. This requires the industrial structure to be adjusted, such as optimizing energy strategy, encouraging public transportation, reducing fossil fuel vehicles, and limiting high energy-consuming industries. At the same time, the population size should be controlled to avoid urban sprawl, and urban greening should be promoted to increase the amount of urban greenery (Fig. 5.2).

5.1.5 Impact of Regional Policy Coordination on Urban Development

Coordinated regional development is of strategic importance to the modernization drive of China. Guided by the pursuit of efficiency and equity, China has introduced a series of policies since the 1980s, such as accelerating the development of the eastern region, developing the western region, revitalizing the old industrial bases in the northeast and promoting the rise of the central region, aiming to form a new landscape where the eastern, central and western regions interact with, reinforce and complement each other with their respective advantages for the purpose of common development. But overall, the landscape of coordinated regional development has failed to take shape. An important reason lies in the fact that the economic development of a region depends largely on the driving force of the central city, whereas

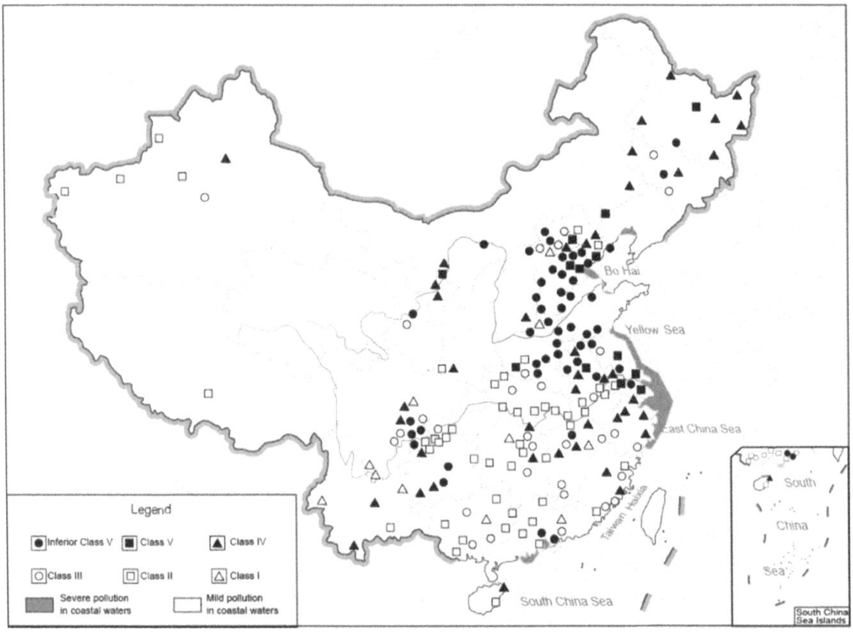

Fig. 5.2 Surface water quality and coastal water pollution in monitored sections across China. *Source* Outline of National Urban System Planning, 2005–2020

there is a lack of linkage between the development policy of the region and that of the central city. Take "the development of the western region" as an example, there is not a clear spatial policy to direct the development at the path to new industrialization, although the guidelines have been put forward—"to take the path to new industrialization characterized by high technology, great economic benefits, low resource consumption and less environmental pollution without damaging the environment; to tap into the advantages in resources, develop high-tech industries selectively and boost tourism and other service businesses in a bid to create a new landscape in the western region featuring economic prosperity, social progress and beautiful scenery." Besides, there is not a clear guidance on urban development that sets out how to manage the resettlement of migrants resulting from environmental protection campaigns that aim to protect the fragile ecosystem in the western region. Environmental protection, industrialization and urbanization are intertwined as part of the systemic issue facing the western region.

Take "the rise of the central region" as another example, there is no systematic analysis and policy guidance on how to balance agricultural industrialization and urban development in the important food producing area, although the guidelines have been put forward—"to make use of the advantages in the location that connects the east and the west, the agricultural industrialization, the abundant land and water resources and the solid foundation for rapid industrial development, to actively explore and expand regional economic and technological cooperation with the east and the west to complement each other's strengths; to boost agriculture and develop the region into an important production base of agricultural products such as grain, cotton and cooking oil; and to foster the pillars of the economy through high-tech transition of traditional industries, so as to play a truly effective role in creating a regional economic landscape where the eastern, central and western regions interact and coordinate with each other for sustainable development". There is no in-depth analysis that centers around central cities in counties on how to find the path to urbanization that reflects the characteristics of densely populated regions with huge labor surpluses. As a result, many places in the central region inevitably follow in the footsteps of the eastern region in attracting investment and building development zones.

In general, due to the vast territory of China, there are huge differences between regions in resource and environment, and the development of different regions is based on varying realities. In the future, the urbanization process and urban development policy will surely be oriented to regional coordination in pursuit of balanced and differentiated development that centers around regional central cities.

5.1.6 Impact of Urbanization and New Rural Construction

Due to the large size of population and high proportion of rural population, China has a strong foundation of traditional agriculture, so the path of China to urbanization is different from the developed Western countries. Even if the level of urbanization reaches 60% or more, there will still be a considerable number of farmers living and working in rural areas. Therefore, an important task for an extended period in the future is to promote agricultural development, accelerate modernization in rural areas and properly resettle the surplus labor by developing medium-sized cities, small cities, and towns. The development and improvement of services in medium-sized cities, small cities, and towns, including agricultural finance, technical services, business services, healthcare, sanitation, and education as well as the processing, circulation, quarantine, testing and export of agricultural products, will play a crucial role in enabling the industry to support agriculture and cities to nurture rural areas. The integration of urban and rural areas is directly reflected in the structure of urban system.

The report of the 16th Party Congress points out that the first two decades of the twenty-first century is a necessary development stage for China to build a moderately prosperous society in all aspects. A well-off society in all aspects means the common prosperity of the whole society, including peasants. Therefore, it is a groundbreaking work to explore the path to sound urbanization with Chinese characteristics that is catering for the realities in China. In a moderately prosperous society, the trend of growing disparities between industry and agriculture, between cities and rural areas and among regions is reversed successfully, which will facilitate the sound and orderly process of urbanization and the improvement of urban development, laying the foundation for building moderately prosperous society in all aspects. The 2004

Strategic Report: China's Sustainable Development quantifies the overall goal of achieving a moderately prosperous society by 2020: the level of urbanization will exceed 55% and the size of urban population will be 700–750 million, and an urban system consisting of large, medium-sized and small cities with a proper structure, complementary functions and maximum efficiency will be created; people's standard of living will be significantly improved, China will have developed from a lower-middle income country to a middle or upper-middle income country, the Engel's coefficient will not exceed 0.35, the Gini coefficient will remain at 0.35 to 0.4, the Human Development Index will be no less than 0.8 and the urban–rural dual structure index will be limited to less than 1.8. The realization of the targets must fully consider the development of rural areas and the coordination and balance between urban and rural areas in terms of functions, facilities, and management.

The above analysis shows that the natural geographical conditions and levels of social and economic development vary greatly from one region to another, so the development approaches of cities must be tailored to fit their realities and create proper spatial organization. The development models of cities must shift from extensive growth to intensive and efficient growth, using all kinds of resources properly. Therefore, to analyze the path to urbanization and urban development from the perspectives of international environment, national land resources, environmental conditions and urban development policies is a sure path to the realization of sound urbanization in China.

5.2 Theory and Method of Urban Spatial Development in China

5.2.1 Solution to the Complex Giant System of National Urban Space

The study of national urban spatial development is a very complex issue. In the past decades, traditional urban system planning summarized the issue into three structures at the methodological level: urban function structure, urban hierarchy structure and urban scale structure. In fact, the emergence and development of cities, which provide space for social and economic development, is a highly complex process. In a general sense, the emergence of a city is directly constrained and influenced by land resources, while the growth of a city follows its own pattern under the influence of such factors as population, industry, and transportation. In the 50 years of urban development in China during the era of planned economy, the growth and decline of urban space were largely influenced by the planned economic system and subject to the planned national industrial space layout. Under the socialist market economy, the factors that influence urban development are more complex. Against the backdrop of economic globalization, the open flow of capital, people, resources, and energy across regions has further diversified the forms of urban space and the mechanism of urban growth.

Therefore, the study of national urban spatial development is a special area that entails the consideration of multiple factors and the planning of various types of spaces. In view of the current theories and planning practices at home and abroad, national urban spatial development is more closely related to land resources planning and regional planning.

Conceptually, land resources planning focuses on the development, utilization, improvement, and protection of land resources [6], while urban system planning is to determine the distribution and development of cities of various population sizes and functions based on regional productivity and division of labor. Although there are different definitions, one thing is certain, urban spatial development is a complex issue involving resources, population, industry, environment, culture, and institutions. Wu Liangyong believed that the city or the region is a complex giant system. [7], and the national urban space consisted of cities and regions is a huge complex giant system. In fact, whether it is population, industry, or ecosystem, any one of the factors is a huge complex system. Therefore, the research on national urban spatial development must establish the theory of method of solving the complex giant systems.

In Introduction to Sciences of Human Settlements, Wu Liangyong proposed the solution to the complex giant system of urban and regional development, including "coherent and integrated research, problem-oriented approach, mastery of the objective law of things like Pao Ding, targeting the crucial point, comprehensive integration and upward spiral [7]". He put forward the five systems of human settlements (nature, human, society, residence, and support) and the five principles for human settlements development (ecology, economy, technology, society, and culture and art). To sum up, we should find the key to the problem and promote coordinated development between cities and regions in a systemic and integrated manner.

From the previous analysis of the theory and practice of urban planning at home and aboard as well as the study of the impact of state intervention on space, we can conclude that in the context of economic globalization, the sound development of national urban space must be based on the guiding principles established by the scientific theory of human settlements; and the theory and method of national urban spatial planning must be built with a broader vision, an approach with higher administrative efficiency and the measures that reflect social equity.

5.2.2 Theoretical Framework

The previous case study of the new round of national urban planning and the analysis of international experience in spatial planning show that national urban planning is an effective means for the central government to regulate national economic and social development within a market economy system. It holds great significance for a country like China that pursues the socialist market economy. The new round of national urban system planning has made tremendous efforts to explore the national urban spatial planning, and its idea of urban system as the core reflects the trait of spatial planning. However, a complete national urban spatial planning should be built from a broader range of fields, particularly analyzing population, resources, and environment.

National urban spatial development planning is essentially a process of formulating and implementing national development policies with certain political significance. In the process, first, the role of macroeconomic regulation and control should be fully embodied in the management of spatial policies and resources to achieve national political goals, for instance, to build "a moderately prosperous society in all aspects". Second, under the market economy, the role of market in resource allocation should be maximized, the global market and resources in the globalized world should be embraced and the free choice of the market should be respected, to strengthen the comprehensive national power in the competition. Third, the ultimate purpose of spatial planning is to improve and enhance the quality of human settlements that should fully consider the construction and planning of urban and rural spaces. Fourth, under the market economy, the disparity between regions is a matter of fact and it is impossible to eradicate disparities completely, but the starting point for development and the public services from the government should be equal, which is the prerequisite for social equity. Therefore, the theory of national urban spatial development planning discussed herein mainly involves the science of human settlements, globalization theory, spatial planning theory, equalization of public services and government intervention.

Science of human settlements: from the perspective of human settlement, a settlement at any level is the integration of five systems: nature, human, society, residence, and support. The development of human settlements at all levels must pursue the integration of five aspects: ecology, economy, technology, society and culture and art. To sum up, no matter how distinctive the extent and level of a space is, the contributing factors and solutions to problems are inseparable from these main influencing factors, which are the key to problem-solving.

Globalization: with cross-border capital flows in the era of globalization, the global production spaces are increasingly decentralized, while the production spaces of certain types of industries are growingly centralized due to the investment behavior of enterprises. Transnational corporations and agglomerations greatly influence the form of global space, and the progress of information technology provides new technical support for the change, so it becomes inevitable that investment flows to developing countries where labor and land costs are relatively low. In such a context, analyzing the trend of foreign investment and grasping the development pattern of industrial space become the important principles of regional spatial organization in the era of globalization.

Spatial planning theory: national spatial planning is the master plan and overarching strategy for continuous coordination on the aspects of economy, society, resources, and environment nationwide, and the arrangement of physical spaces to allocate major resources and other factors of production in regional economic activities on a broad scale. Through planning, the market economy of the country is improved and becomes more competitive, to avoid and correct the market failures under a completely free market economy. In the administrative system, the planning process coordinates the interests of various departments, the central government, and local governments, which guarantees the mutual interests for all.

Equalization of public services: economic disparity is only the surface of inequality in development, and what lies beneath is the deeper problem of inequality in political power and the lack of development capability. Equal opportunity means to start on a level playing field, namely the same starting point for economic competition. The government plays a leading role in promoting social equity in livelihood, health, and education, narrowing wealth gaps, and supporting vulnerable groups, which is essential to social equity and coordinated regional development (Fig. 5.3).

Government intervention: according to the theory of circular and cumulative causation, the market forces have the tendency to increase regional disparities, rather reducing them, which necessitates government intervention in the economy to promote the coordinated development of regional economies. At the same time, in a completely market-driven environment, the competition among local governments is inclined to engender moral hazards, so the coercive power or influence of a third party (government at higher levels) can resolve conflicts that would otherwise not be resolved through negotiation, thus turning non-cooperative game into cooperative game.

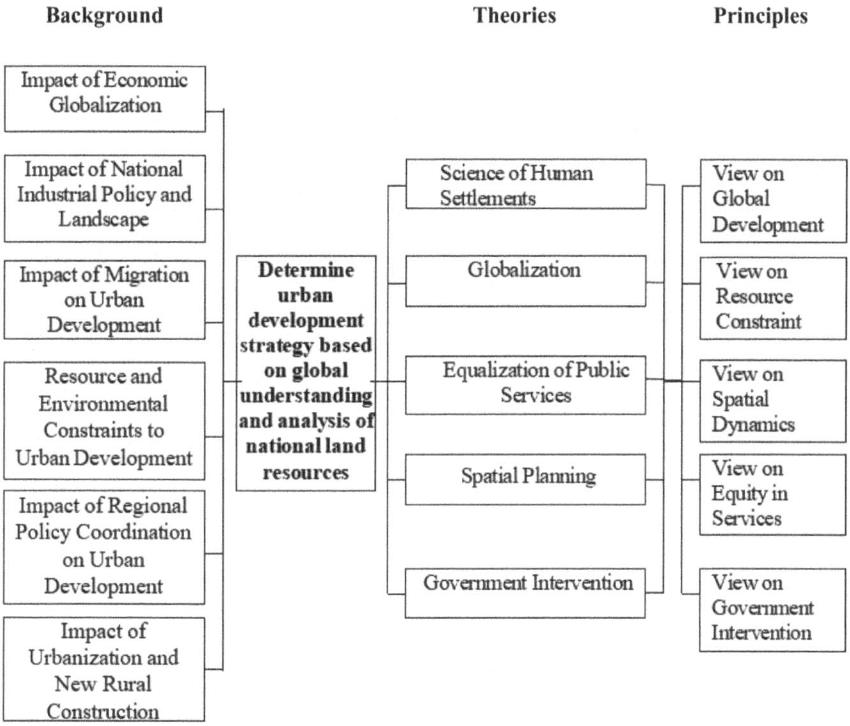

Fig. 5.3 Theoretical framework of national urban spatial planning research

Based on the aforementioned theories, the national urban spatial development planning theory should establish five fundamental concepts:

(1) Global Development Perspective

The national spatial strategy must be tied to global economic and social development. To enable cities in different regions to integrate into the world economic trend, make effective use of international markets and international capital, take full advantage of international resources (energy, technology, culture, management) and promote national economic development; to build a globally oriented spatial structure (global network, node cities and regions), integrate into the world city system and support the development of mega-city regions, gateway cities and regions with high concentrations of economic resources.

(2) Resource Constraint Perspective

To focus on the constraints of resources and environment on spatial development and take the path of sustainable development. Guided by the science of human settlements, to integrate the five systems—nature, human, society, residence, and support, to determine the master plan and specific policies for human settlement development from the perspective of the scientific and effective use of spatial resources, and to develop a spatial system characterized by ecological security, economic efficiency, and urban–rural coordination.

(3) Spatial Dynamics Perspective

To embrace the trend of global development, focus on the development priorities at the current stage and determine an open and flexible urban spatial structure. To create an analysis system for the correlation between "economic space of flows" and "human settlement space", enhance the resilience of spatial structure, monitor the trend of regional spatial development, and adjust the priorities of spatial development in a timely manner. To improve the services of regional transportation infrastructure and provide favorable conditions for spatial growth.

(4) Equity in Services Perspective

To build a standard system of public service facilities that integrates urban and rural areas under the principle of equity in services. To lay down the requirements for establishing public service facilities of education, health, culture and so on, and propose supportive policies for spatial development, especially financial and tax policies. Equity between regions should be more reflected in the equalization of public services.

(5) Government Intervention Perspective

The central government intervenes in the socioeconomic development of the entire country at the macro level through national spatial planning, remedies the shortcomings of market through macroeconomic regulation, reflects the political belief of fairness, deals with the intricate relationship between the central and local authorities and among local authorities, determines policy interventions for spatial development at different levels, protects precious resources that have a bearing on national interests and the safeguarding of the quality of overall human settlements, and guarantees the coordinated development of economy, society and environment.

5.2.3 Planning Methodology

The preceding analysis shows that the traditional approach to land resources planning focuses on the management and arrangement of land resources for the sake of productivity, while the analysis of urban spatial development is not deep enough. The traditional approach places great emphasis on building three systems—functional structure, scale structure and hierarchical structure, with insufficient emphasis on resource conservation and management. To fill the gap, a new direction for the development of urban planning theory and method has emerged, that is, to establish a comprehensive urban spatial development planning system. In recent years, the widely advocated spatial planning in Europe represents a significant innovation in urban and rural planning, the core of which lies in the comprehensive analysis of the needs and possibilities of urban development from a broader range of perspectives including regional planning theory and urban planning theory as well as economics, political science and ecology related to urban development. Thereupon, a city-centered spatial system is built that not only meets the requirements of urban development, but also uses the concept and method of sustainable development as a prerequisite and constraint of urban development to realize integrated and coordinated development (Fig. 5.4).

In the context of the increasingly market-based economic system, mounting pressure on resources and environment and accelerating rate of urbanization, it is necessary to establish an all-inclusive spatial planning system based on the comprehensive analysis of urban development from the perspectives of population, resource, industry, environment, culture, and institution, which is a response to the call of our era and required for the expansion of the scope of urban planning.

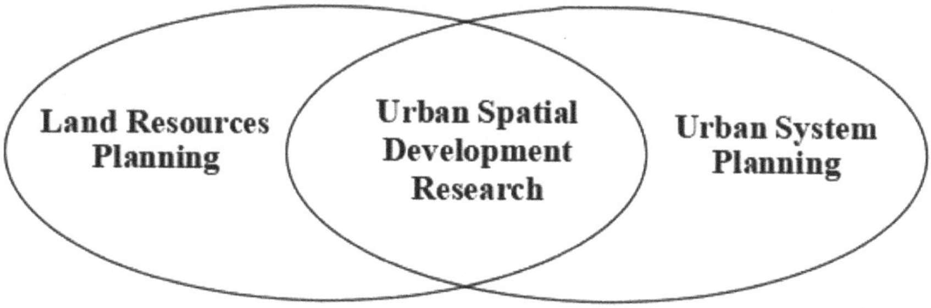

Fig. 5.4 Urban spatial development research based on land resources analysis

5.3 Technical Framework of Urban Spatial Development Research in China

The previous theoretical overview, historical evolution and case study provide technical support for developing the urban spatial planning suited to the national conditions of China. According to the above analysis, the technical roadmap of spatial planning in China is summarized as follows: to analyze the position of China in the global economic system and political landscape in the context of globalization, build a spatial system that is closely integrated with the world, especially the Asia–Pacific region, identify the priority regions and enhance the country's overall competitiveness; to carry out spatial zoning based on ecological network and establish the policy framework for spatial resources conservation and utilization; to arrange the distribution of industries and population based on the analysis of population, industry and space; and to establish the comprehensive planning management system to coordinate spatial development in a holistic manner.

5.3.1 Identify Cities and Regions with International Competitiveness in Response to the Trend of Economic Globalization

5.3.1.1 Analyze International Capital Flows and Gain Insight into the Development Trend of Priority Cities and Regions

A thorough analysis of the interactions between capital flows, industrial linkage, regional collaboration, and urban space in China in the context of globalization is the basis for the analysis of the country's urban spatial structure. Since 1990, the inflow of foreign capital into China has increased year by year, and China has become the only developing country in the world that has significantly increased foreign capital. According to the definition of a transnational corporations by the UN agency, most of the foreign invested enterprises in China are transnational corporations, except those invested by natural persons from abroad. Therefore, the distribution of foreign investments in China reflects, to a considerable extent, the distribution of transnational corporations' investments in China. The analysis of foreign investment represented by transnational corporations helps us to understand the changing trend of regional space. It has been argued that transnational corporations' investment in China has gone through three stages [8, 9]: the first stage (1979–1992), predominated by commodity trade, technology trade and few tentative direct investment; the second stage (1992–2001), the start of large-scale direct investment, mainly in the manufacturing sector; and the third stage (after 2001) when transnational corporations fully entered the Chinese market, including the service sector.

In terms of industrial structure, foreign investment is mostly concentrated in the secondary sector. As of 2001, the number of foreign investment projects in the secondary sector accounted for more than 73% of all foreign investment projects. Most of the 2519 projects invested by the Fortune 500 companies in China are in the secondary sector (Table 5.2).

In terms of the influence of transnational corporations on regional space, foreign investment is mostly concentrated in the eastern region of China. According to statistics, as of 2001, the projects of foreign direct investment in the eastern region accounted for 81% of all approved projects, and the actual utilization of foreign investment in the eastern region accounted for 86%. According to the information available, the majority of the 2519 projects invested by Fortune 500 companies in China are also concentrated in the eastern region [9] (Tables 5.3 and 5.4).

A study has analyzed the product mix and geographical distribution of "Made in China" products with global influence, showing that by the end of 2002, China ranked first in the world in terms of the production of 172 manufacturing products,

Table 5.2 Industrial structure of foreign direct investment as of 2001

Sector	No. of projects	Proportion %	Contractual foreign investment	Proportion
Total	390,025	100	7452.91	100
Primary	11,242	2.88	140.71	1.89
Secondary	285,000	73.07	4631.60	62.14
Tertiary	93,783	24.05	2,680.60	35.97

Source China Foreign Investment Statistics 2002, former Ministry of Foreign Trade and Economic Cooperation

Table 5.3 Regional distribution of foreign investment in China as of 2001 unit of foreign investment: 100 million USD

Region	No. of projects	Proportion (%)	Contractual foreign investment	Proportion (%)	Actual utilization	Proportion (%)
Total	390,025	100	7,452.91	100	3,952.23	100
Eastern	315,053	80.78	6,439.23	86.40	3,397.26	85.96
Central	46,713	11.98	565.21	7.53	346.93	8.78
Western	28,259	7.24	448.47	6.02	208.04	5.26

Source China Foreign Investment Statistics 2002, former Ministry of Foreign Trade and Economic Cooperation

Table 5.4 Industries invested by fortune 500 companies in China

Industry	Manufacturing	Service	Business	Others
Quantity	773	185	76	21
Proportion %	73.27	17.54	7.20	1.99

Source China Foreign Investment Statistics 2002, former Ministry of Foreign Trade and Economic Cooperation

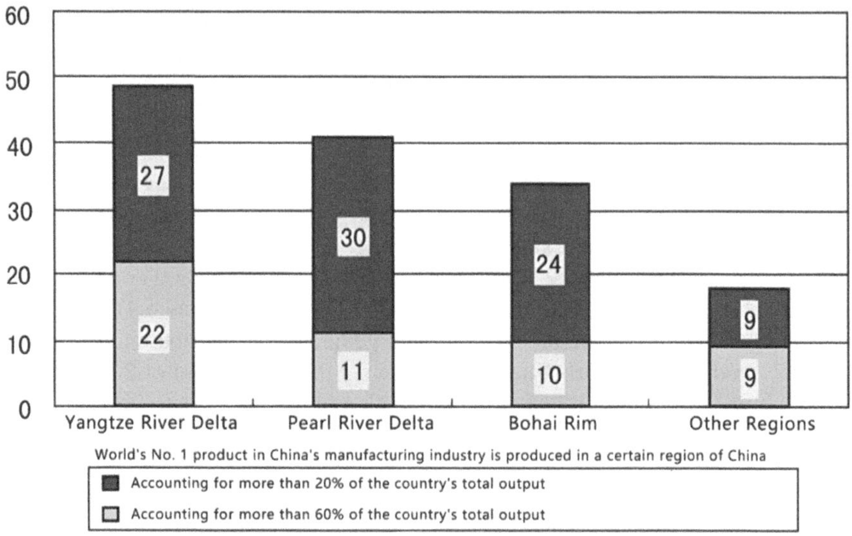

Fig. 5.5 Distribution of main production areas in China's manufacturing industry. *Source* Great Wall Enterprise Institute, Study of World's Top "Made in China" Products, 2004, 1

covering 28 fields, with most of their production concentrated in the eastern coastal region (Fig. 5.5).

The above figures and tables clearly illustrate the geographical characteristics of China as the "workshop of the world", the differences in the spatial structure of cities in the eastern and western regions, and the underlying reason for the huge wave of labor migration.

In the new century, there is a new trend in foreign investment led by transnational corporations in China. First, the investment has been strengthened. In 2001, China approved 69.2 billion USD in new contractual foreign investment, up 10.43% year-on-year; in 2002, China approved 82.8 billion USD in new contractual foreign investment, up 19.62% year-on-year; in 2003, China approved 115.070 billion USD in new contractual foreign investment, up 39.02% year-on-year; and in 2004, China approved 153.5 billion USD in new contractual foreign investment, up 33.38% year-on-year. The total amount of foreign investment in China has now reached 1096.6 billion USD, and foreign investment remains the main driver of China's economic growth. Second, investment projects have been systematized. Investments in manufacturing have continued, so that China remains the "workshop of the world", especially in the upstream of manufacturing, such as large petrochemical operations concentrated in Jiangsu, Zhejiang, and Guangdong; investments in R&D have been active, turning China into a regional R&D center, such as the R&D centers of Microsoft and Nokia in Beijing, Shanghai and other central cities; and investments in production services have begun, turning China into a regional operations management center. Third, investments are more concentrated in certain regions. Petrochemical and electronics projects are generally located in coastal port areas and manufacturing hubs, and service projects such as R&D and financial circulation are concentrated in the coastal cities, mainly in the Yangtze River Delta, Pearl River Delta, Beijing-Tianjin-Hebei region, and Bohai Economic Rim. As of 2004, 85.6% of foreign investment was in the eastern region and 4.9% in the western region. In recent years, the investment in the eastern region has accounted for over 82%, and there is no tendency of large-scale shift to the central and western regions [9].

Under this trend, the urban spatial structure of China will continue to feature the increasing concentration of cities in the eastern coastal areas, especially the Yangtze River Delta, Pearl River Delta, Beijing-Tianjin-Hebei region, and Bohai Economic Rim where the dense urban clusters will continue to expand. Different forms of regional urban space will emerge, including megacity, metropolitan area, and city cluster, which will form a network structure together with medium-sized and small cities and towns. The central and western regions will still be characterized by the development of central cities, and medium-sized and small cities. In view of this, the focus of national spatial policy should be.

First, to establish the spatial policy for Mega-City Region in coastal areas as soon as possible. The renewal of urban functions in coastal areas will be enhanced to build the internationally oriented transportation and information infrastructure and improve the country's ability to participate in global competition. Attention will be paid to the demands of large industries on space and their impact on environment. The internal spatial structure of cities will be adjusted based on the development trend of production services. The regional urban structure will be decentralized at a faster pace to alleviate the pressure from dense structures and environment degradation. At the management level, the planning and management mechanism for the Mega-City Regions centering around Shanghai, Beijing, and Guangzhou will be established to guarantee the overall integrity and environmental quality of those regions.

Second, to accelerate the development of transportation links in Mega-City Regions to enlarge the scope of their influence, create conditions for the extension of transnational corporations' value chains and the shift of traditional industries from central cities in the regions, and give priority to the development of cities along the transportation axis. Take the Yangtze River Delta as an example, a greater economic zone that is not confined to the 15 cities in the Yangtze River Delta and includes eastern Anhui, northern Jiangxi, and northern Jiangsu and southern Zhejiang with Shanghai as the core should be built as soon as possible to promote coordinated development of the region.

Third, to enhance the supply of infrastructure and actively foster the development of central cities in the central and western regions to attract investment. Take the southwestern region for example, based on the excellent technical foundation and abundant labor supplies, Chongqing and Chengdu will be cultivated into the bridgehead for attracting foreign investment in the region, thereupon driving the development of Sichuan, Yunnan, and Guizhou.

5.3.1.2 Analyze the Impact of Regional Economic Cooperation Organizations and Enhance the Development of Node Cities and Cross-Border Transit Routes

In recent years, China has established several regional organizations together with neighboring countries. The establishment of the urban spatial structure in China, a country located in the Asia–Pacific region, should be largely aimed at the common development of the Asia–Pacific region.

The major regional organizations that China has participated in include Asia–Pacific Economic Cooperation, ASEAN, and Great Mekong Subregion Cooperation. On the macro level, the external economic links, transportation routes, and information exchange of China's urban space are mainly oriented toward these regions. The major impacts on the overall urban landscape include: first, the core urban areas will be connected with the above regions in terms of economic management, information exchange, import and export of staple commodities; second, core cities in the border areas will become the nodes that are linked up with the above regions, and economic zones in the borders areas with the nodes as the core will be developed, and this trend has already been observed in Nanning, Kunming, Harbin and Urumqi; third, the development of routes for cross-border resource management and energy import and export as well as the management of regions along the basins and transportation links will become important elements of spatial management. For that reason, the spatial policy should:

First, focus on the development of cross-border transit routes and the building of a transport corridor gearing towards the Asia–Pacific region. Second, cultivate new growth poles based on the core cities in the border areas and establish cross-border economic cooperation zones in the northeast with Harbin as the center, in the southwest with Nanning and Kunming as the center, and in the northwest with Urumqi as the center, respectively, to promote regional economic development. Third, give priority to the development along the oil import corridors covering Russia, Kazakhstan, the Middle East, and southeast Asia; take a holistic approach to the common development and management of water resources in the Mekong River basin; and strengthen the development of logistics ports based on small and medium-sized border cities.

5.3.1.3 Fully Recognize the Role of Industrial Cluster and Guide the Operation of Small—and Medium-Sized Cities and Small Towns by Taking Advantage of the Situation

With the advancement of market-based reform, the role of industrial cluster is widely recognized. It has been argued that industrial clusters are compatible with the global shift of manufacturing and can give full play to the fundamental regulatory role of market mechanism, promoting the coordinated development of large, medium, and small cities, and between urban and rural areas [2]. In China, industrial clusters are quite common at the local level, such as the electronic information industry in Zhongguancun of Beijing, the information industrial cluster on the east bank of the Pearl River, the cashmere industry in Qinghe of Hebei, the footwear industry in Jinjiang of Fujian, the motorcycle manufacturing industry in Chongqing, the domestic appliance industry in Qingdao of Shandong, and the fireworks manufacturing industry in Liuyang of Hunan. In Zhejiang and Guangdong, industrial clusters have a particularly prominent impact on regional space. In the Pearl River Delta, One-fourth of the 404 towns are characterized by industrial clusters. The information industrial cluster on the east bank of the Pearl River includes dozens of towns in Guangzhou, Dongguan, Huizhou, and Shenzhen, worth 300 billion CNY. The electrical machinery industrial cluster on the west bank of the Pearl River includes a dozen of towns in Shunde, Zhongshan, Nanhai, Jiangmen, Zhuhai, and Guangzhou, worth 130 billion CNY. In Zhejiang, there are 118 clusters of private enterprise with an annual output of 1 to 5 billion CNY, 26 clusters with an annual output of 5 to 10 billion CNY and 3 clusters with an annual output of over 10 billion CNY, of which the output of 53 clusters accounts for more than 30% of the total in the domestic market. The production of windproof lighters in Wenzhou accounts for 70% of the world's total and lighter manufacturers are easily found in the large and small towns in Wenzhou [10]. The clusters have formed local economic masses and created regional economic space featuring continuous self-improvement and growth.

Given the pace of economic globalization and the huge agricultural population in China, boosting the development of industrial clusters featuring specialty industry is not only an important means to embrace the global shift of industries, but also a critical measure to enhance the relevance of township and village enterprises. The prosperity of medium and small cities and small towns in Guangdong and Zhejiang is largely due to the flourishing industrial clusters of various kinds and the extensive economic engagement with the world brought about by the industrial clusters. The fundamental reason for the underdevelopment of medium and small cities and small towns in China in the past decade is the sluggish urban economy, which has not only affected the development of medium-sized and small cities and small towns, but also driven the movement of surplus labor to major cities and megacities, resulting in the overdevelopment of major cities and underdevelopment of small cities and towns. To cultivate the core competitiveness of small—and medium-sized cities and re-create the development momentum for small towns is an important part in achieving sound urbanization and promoting the coordinated development of urban and rural areas. The coordinated development of cities and different sizes and small towns are not just words, but a development path backed up with substance, in terms of both economy and urbanization pattern analysis (Fig. 5.6).

All in all, it requires the creation of an open and sparse spatial structure on the national level, integration into the world economy and common development in the Asia–Pacific region in the context of economic globalization and regional economic integration. With a coordinated approach to the industrial and spatial organization in urban and rural areas, the coordinated development of urban and rural areas is promoted.

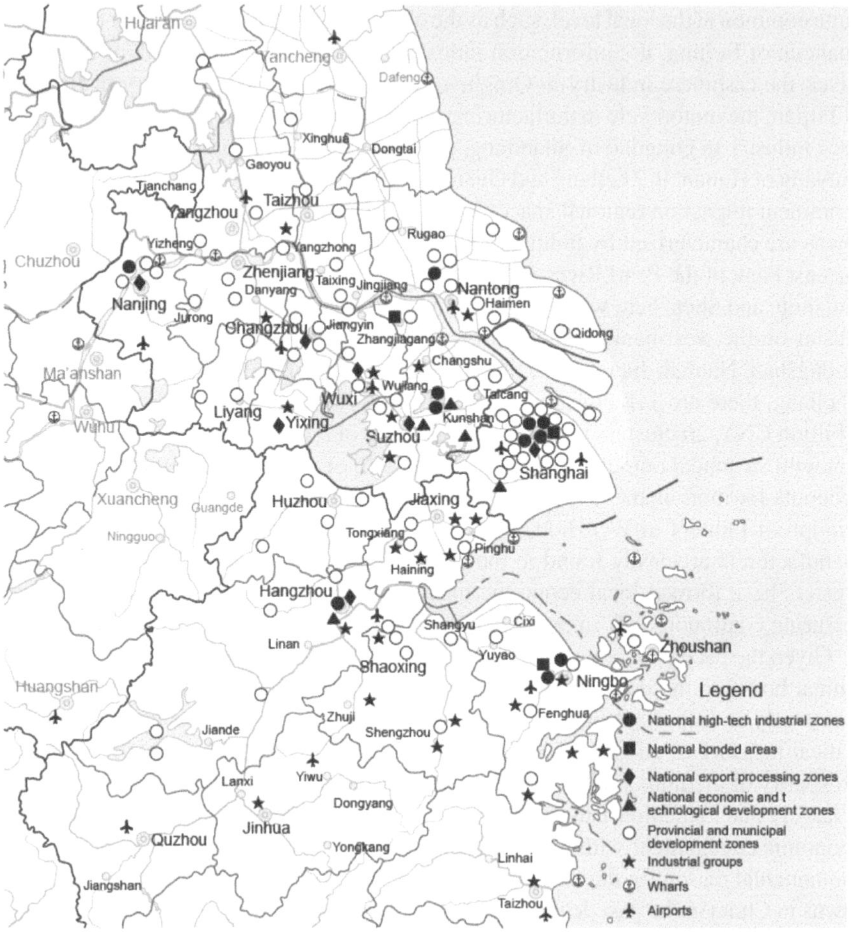

Fig. 5.6 Industrial clusters and urban spatial distribution across the yangtze river delta

5.3.2 Determine the Protection and Management of Precious Resources at Different Levels Based on Sustainable Urban Development

5.3.2.1 Analyze the Layers of Spatial Resources Based on Ecological Security

The spatial resources are divided into ecological security layer, infrastructure layer and human settlement layer, and the analysis of the three layers is used as the basis for constructing a planning method for the sound development of urban space. The ecological security layer refers to the spatial resources based on natural and ecological elements, mainly water bodies, forests, wetlands, grasslands, biologically diverse areas, national nature reserves and scenic spots, as well as terrain and geomorphologically restricted areas (mountains, snowy plateaus, sandy areas, Gobi, alkali soil, swamps, glaciers and permanent snow) and erosion prone areas (slope > 25 degrees), which have respective significances at national, regional and city levels. The infrastructure layer refers to the regional and urban transportation infrastructure underpinned by railroads, highways and airports, digital transit network and other municipal infrastructure required for urban security, which support and promote the development of cities and regions and constitute the backbone of the space. The human settlement layer refers to the human settlements of various types and at different levels on which human lives revolve, such as cities, towns, villages, and the living quarters around industrial mines, where intricate relationships exist between cities and rural areas, among cities and among rural areas. In any case, the three types of spaces with distinct functions are clearly delineated, and the analysis of spatial resources at the national, regional and city levels is a crucial prerequisite for sound planning (Table 5.5).

The planning should also adhere to the principles of environmental protection and put forward specific requirements on the siting of urban development, prevention of soil erosion, and protection of ecologically sensitive areas. In terms of environmental pollution control, it is proposed to develop circular economy, establish green national economic accounting system, and reduce the requirements of total pollutant emissions and emission intensity. It is proposed that urban development should be based on the environmental carrying capacity (atmospheric environmental capacity, water pollution absorption capacity), and the possible impact of urban development on the ecological environment should be fully considered.

It is particularly noteworthy that as a country where the per capita arable land is only one fourth of the world average, the importance of arable land protection cannot be overemphasized. It makes clear that 1.1 million km2 of basic cropland is the bottom line for food security. Given the fact that areas suitable for urban development are often the main food-producing areas, we should focus on balancing the relationship between urban and rural development and arable land protection in the regions and specify the basic requirements for land conservation during urban development.

As the southern and eastern regions are water-rich while the northern and western regions are water-stressed, urban development should consider the regional distribution of water resources, especially the status quo of basins, to properly determine the layout and scale of cities. For the current inter-basin water transfer projects in China, such as the South-to-North Water Diversion Project, the impact of water transfer on ecological environment should be considered. The long-term solution to water stress in urban development is the change of urban growth model, rather than inter-basin transfer.

As China is facing energy shortage, urban development should not only reduce energy consumption, but also explore the use of new and renewable energy, which should be taken seriously especially by resource-rich regions. In the next 20 years,

Table 5.5 Three layers of ecological elements of the ecological security landscape

National level	Areas with important ecosystem services	Large rivers, large lakes, headwaters, important water conservation areas, important forests, large contiguous wetlands, important grasslands, important desert areas, biologically diverse areas, national nature reserves, scenic spots, and forest parks
	Ecologically fragile areas	Geomorphologically restricted areas (large undulating mountains, snowy plateau, sandy areas, Gobi, alkali soil, swamps, glaciers, and permanent snow), low-precipitation areas (average annual precipitation < 50 mm), erosion prone areas (slope > 25 degrees), areas with serious water erosion, large contiguous areas with land desertification, areas with strong soil erosion, contiguous stony deserts
Provincial level	Areas with important ecosystem services	Rivers, lakes, water conservation areas, headwater conservation areas, reservoirs, wetlands, woodlands, grasslands, biologically diverse areas, national, provincial, and municipal nature reserves, scenic spots, forest parks, flood storage areas, basic cropland protection areas, green corridors, and ecological isolation zones
	Ecologically fragile areas	Geomorphologically restricted areas (large undulating mountains, snowy plateau, sandy areas, Gobi, alkali soil, swamps, glaciers, and permanent snow), low-precipitation areas (average annual precipitation < 50 mm), areas with frequent geological disasters (earthquakes, mudslides, floods, droughts), erosion prone areas (slope > 25 degrees), areas with serious water erosion, areas with land desertification, areas with strong soil erosion, stony deserts, and areas with serious groundwater overdraft
City level	Areas with important ecosystem services	Rivers, lakes, water conservation areas, headwater conservation areas, reservoirs, wetlands, woodlands, grasslands, biologically diverse areas, national, provincial, and municipal nature reserves, scenic spots, forest parks, flood storage areas, basic cropland protection areas, green corridors, urban green line control area, urban green space and greenery along railroads and trunk roads as part of green corridor
	Ecologically fragile areas	Areas with frequent geological disasters (earthquakes, mudslides, floods, droughts), erosion prone areas (slope > 25 degrees), areas with serious water erosion, areas with land desertification, areas with strong soil erosion, stony deserts, sinkholes in coal mining areas, airport noise control areas, major municipal transportation route control areas and areas with serious groundwater overdraft

China will have a great opportunity to make use of overseas oil and gas resources. Since obtaining cost-effective and stable oil and gas resources from abroad is an inevitable trend, it is crucial to enhance the development of cities based on oil ports in the coastal areas, the northeast, and the northwest.

5.3.2.2 Establish a Three-Level Spatial Resource Planning and Management System

In the broad sense of space, cities and regions have "space of flows" characterized by economy, such as capital space and production space, and "fixed space" characterized by nature, culture, and human settlement, such as mountains, rivers, historic sites, and towns. The former is fluid, active and changing, while the latter remains unchanged despite the passing of the years in history. Friedman argues that there have always been two properties of space, and in the era of globalization, economic space has been disrupting, invading, and disintegrating the living space of individuals and societies, a trend that must be curbed [11]. Therefore, establishing a three-level planning and management system for national, regional, and urban planning with resource conservation (natural and cultural) as the main content based on the three-level analysis of spatial resources is an important prerequisite for creating a sound plan (Table 5.6).

In terms of resource protection, the national-level spatial planning should focus on the following two aspects. First, natural resources: water resource areas, including major drinking water sources of national importance, interprovincial drinking water

Table 5.6 Content of spatial planning system and resource management

National spatial planning	Natural resources: large rivers, large lakes, headwaters, important water conservation areas, forests, large wetlands, grasslands, national nature reserves, scenic spots, and forest parks and geoparks	Cultural resources: World Cultural Heritage; national-level historic and cultural cities, towns, and villages; cultural relics under national protection
Provincial spatial planning	Natural resources: inter-provincial rivers, lakes, water conservation areas, wetlands, woodlands, grasslands, provincial nature reserves, scenic spots, flood storage areas, basic cropland protection areas, green corridors, and ecological isolation zones	Cultural resources: major provincial-level historic and cultural sites and relics; historic cities, towns, and villages; cultural relics under provincial protection
Urban master planning	Natural resources: important rivers in cities, lakes, water conservation areas, wetlands, woodlands, grasslands, flood storage areas, basic cropland protection areas, urban green line control area, green space and greenery along railroads and trunk roads	Cultural resources: city-level historic and cultural sites and relics; spatial layout and important buildings of historic and cultural cities; historic towns and village; cultural relics under the city authority's protection

sources, major rivers, lakes, reservoirs and other important drinking water sources, as well as the conservation areas of the water sources; forest resource areas, including Daxing'an and Xiaoxing'an mountains, Changbai Mountain Range, Sichuan and Yunnan, southeastern Xizang, and Altay; wetland areas, including Sanjiang Plain in the northeast, the Yellow River delta, the coast of northern Jiangsu, the Ruoerge plateau, the Yarlung Tsangpo River, Lhasa River and Nianchu River in Xizang, and dry-hot valleys in Sichuan and Yunnan; and the areas prioritized for ecological restoration, including grasslands in Inner Mongolia, areas with land desertification in the northwest, areas with severe soil and water erosion on the Loess Plateau, areas with soil erosion in the Dabie Mountains and Karst stony deserts in the southwest. Second, historical and cultural resources: regional historical and cultural resources, such as those along the Yellow River, the Yangtze River and the ancient Silk Road; historical and cultural resources of towns and cities, such as national-level historic and cultural cities and several national-level historic and cultural towns and villages; and protected historical and cultural sites, such as cultural relics under the protection of national, provincial and city authorities.

5.3.3 Scientific Prediction of Population Growth and Movement and Accurate Identification of Priority Areas for Urbanization

5.3.3.1 Establish a Gray Prediction Model of Industrial Development, Population Movement and Spatial Expansion

The evolution of urban spatial structure is a process of continuous adjustment amid population movement and industrial development. In this process, people are the most active factor, which have been constantly changing with the migration of jobs. Over the past 20 years, this change has been reflected in the large-scale movements of migrant labor force across regions in China. According to the Fifth National Population Census in 2000, there were 57.45 million non-registered permanent residents in the urban areas who had migrated from the rural areas, raising the level of urbanization in the country by about 4.6%. According to the analysis of a sample of 1% population, the agricultural population absorbed by the cities in China is about 53 million, and that absorbed by the towns is around 23 million, totaling 76 million, which demonstrates that the flow of rural surplus labor to urban areas is an important factor in accelerating the urbanization process in China.

According to the projections of the former National Family Planning Commission, the total population of China will reach a peak of 1.5 billion by 2033. Until then, the total population will continue to increase at 8–10 million people per year, and what is significant is that the working-age population, the total population, and the population aged 65 and over will hit their peaks. In the future, the scale of population movement and migration will increase, and population movement will continue to follow the basic pattern of migrating from rural areas to urban areas, from less developed regions and developed regions, and from small to medium-sized cities to larger cities. metropolitan interlocking region, urban clusters and regional central cities will be the main space to absorb the floating population. In terms of the destination of population movement, the coastal areas will still be the first choice for the floating population before 2015, and the proportion of the floating population absorbed by the coastal areas will stabilize and decrease after 2015. In the central region, however, the inflows and outflows in the will reach an equilibrium around 2015 because of economic development, ending the trend of net outflows.

However, the fundamental reason for the large-scale movements of migrant labor force across regions is that the destinations can provide large numbers of jobs, while the provision of jobs depends on the development of industries. If the prediction of population movements is not based on the background analysis of industrial development, its role in guiding spatial planning would be greatly restricted. Therefore, it is crucial to conduct the analysis of industrial development, population movement and spatial growth. Regarding the current level of industrialization in China and the industrial structure of the destinations, the key factor influencing the development of secondary industry is the investment of capital. Therefore, the urban space as a system entails complex correlations between spatial expansion and capital, industry, and population. But the city is a complex giant system with plenty of information and working mechanisms unknown to us, so it's difficult to accurately describe the systematic relations and structures which can only be expounded with logical reasoning and general principles. Hence, the establishment of a gray prediction model based on capital, industry and population is an important prerequisite for understanding the growth of urban space.

5.3.3.2 Identify Priority Areas for Urbanization Based on the General Trend of Population Movement

The population density in China is mainly influenced by the geographical conditions and the level of economic development. Plains and basins are the primary densely populated areas, mainly in North China, Northeast China, the Han River, central Shaanxi, Chengdu, Chongqing, middle and lower reaches of the Yangtze River, the Pearl River Delta, and coastal plains. With the development of regional economy and the widening of urban–rural disparities, the gradual movement of population to economically developed regions and urban areas remains an inevitable trend. And with the growing economic influence of central cities, the development of urban clusters around central cities is gaining a positive momentum. The status of population concentration showed that the Pearl River Delta, the Yangtze River Delta, and Beijing-Tianjin-Hebei region are the secondary densely populated areas. Given the trend of future development, the regions with faster economic growth, including south-central Liaoning, Shandong Peninsula, southeast Fujian, Wuhan, central Shaanxi (Xi'an), Central Plains (Zhengzhou), Changsha-Zhuzhou-Xiangtan,

Chengdu-Chongqing, Changchun-Jilin, and Beibu Gulf, will see high concentrations of the floating population, and thus become prioritized for urbanization.

In addition, according to the analysis of the Fifth National Population Census, the inflows into large cities of which the populations exceed 500,000 account for 71.1% of the national total, the inflows into cities of which the populations exceed 200,000 account for 93.2% of the national total. Therefore, cities of medium size and above are the main absorbers of floating population in China. In other words, prefecture-level cities and developed county-level cities are the main space to accommodate the floating population in China, which indicates that more large and medium-sized cities will be needed to further increase the level of urbanization in the future. At the same time, due to China's special national conditions, the development of small cities is restricted by the levels of administrative divisions, and the total size of urban expansion has been controlled for many years, so the small cities fail to play a proper role in the urban system. How to give full play to the role of small cities and towns in absorbing the floating population is a key issue to address in the future process of urbanization, which, however, is a major policy issue that cannot be properly dealt with simply on the technical level (Fig. 5.7).

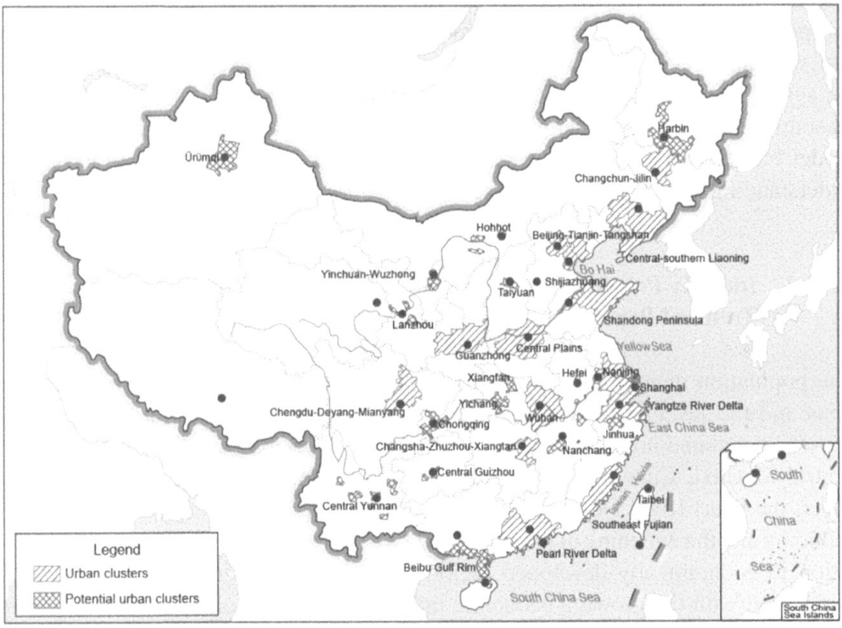

Fig. 5.7 Existing and potential urban clusters in China

5.3.4 Establish a New Regional Coordination Mechanism for Coordinated Regional Development

5.3.4.1 Establish Urbanization Policy Zoning Centering on Regional Central Cities

Regional policy is one of the core elements of national spatial planning. Since the 1980s, China has established different regional development policies for the eastern, central, western, and northeastern regions. On the basis, the new round of national urban system planning determined the urbanization policy zoning. In fact, urbanization policy zoning is significant not only in a political sense, but also in terms of resources, and even more important in the sense that the regional policies can boost and drive the overall socioeconomic development of the country. Since the Reform and Opening Up, China has introduced a series of regional policies, from the trailblazing development of the eastern region, the large-scale development of the western region, the revitalization of the old industrial bases in the northeast to the rise of the central region, which should be regarded as expedient policies considering their context. Today, given the current development in the eastern, central, and western regions, it is necessary to propose a new regional policy zoning that is both in line with the overall interests and suited to local social and economic conditions.

Historically, China has introduced different versions of regional policies for various purposes. Some were developed from the perspectives of regional policy, regional economic development, and urban economic zones. The policies that have been developed from the perspective of national policy include:

Six Economic Cooperation Zones: In 1954, China established seven economic cooperation zones—the Eastern Zone, Northeastern Zone, Northern Zone, Central Zone, Southern Zone, Southwestern Zone, and Northwestern Zone. In 1961, the Central Zone and Southern Zone were merged into the South-Central Zone, resulting in six Economic Cooperation Zones nationwide. Administration and war preparation were the main considerations of the first regional economic policy in the country (Table 5.7).

Table 5.7 Six economic cooperation zones

Economic zone	Scope
Eastern zone	Shandong, Jiangsu, Shanghai, Zhejiang, Fujian, Anhui, Jiangxi
Northern zone	Beijing, Tianjin, Hebei, Shanxi, Inner Mongolia
South-central zone	Henan, Hubei, Hunan, Guangdong, Guangxi
Northeastern zone	Heilongjiang, Jilin, Liaoning, eastern Inner Mongolia
Northwestern zone	Shaanxi, Gansu, Ningxia, Qinghai, Xinjiang
Southwestern zone	Sichuan, Yunnan, Guizhou, Xizang

The Three Major Economic Zones: During the Seventh Five-Year Plan, the central government divided the country into Three Major Economic Zones: the Eastern Zone, Central Zone, and Western Zone based on administrative divisions, which had a significant impact on subsequent policies and institutions. However, the division of the three zones only reflected the descending levels of economic development from the east to the west and failed to enable the core cities to drive regional development in coordination (Table 5.8).

The Seven Economic Regions: In the Ninth Five-Year Plan and Long-Range Objectives to the Year 2010, the government proposed to gradually develop seven cross-provincial and cross-regional economic zones according to the laws of market economy, internal economic ties and natural geographical features and underpinned by central cities and major transportation routes. The proposal placed more emphasis on the economic ties between provinces and regions but was inconsistent with the administrative divisions based on geographic areas in the absence of scientific analysis of internal economic ties between regions (Table 5.9).

The Concept of Eight Comprehensive Economic Zones: During the Eleventh Five-Year Plan, the think tank of the State Council released a report on "Strategies and Policies for Coordinated Regional Development", which proposed to divide the country into the eastern, central, western, and northeastern regions and on that

Table 5.8 The three major economic zones

Economic zone	Scope
Eastern zone	Beijing, Tianjin, Hebei, Shandong, Jiangsu, Shanghai, Zhejiang, Fujian, Guangdong, Guangxi, Hainan
Central zone	Heilongjiang, Jilin, Inner Mongolia, Shanxi, Henan, Hubei, Hunan, Anhui, Jiangxi
Western zone	Xinjiang, Qinghai, Xizang, Gansu, Ningxia, Shaanxi, Sichuan, Guizhou, Yunnan

Table 5.9 Ninth five-year plan

Economic zone	Scope
Yangtze river delta and riverside economic region	Shanghai, Jiangsu, Zhejiang and riverside areas in Anhui, Jiangxi, Hubei, Hunan, and Sichuan
Bohai rim region	Beijing, Tianjin, Hebei, Shandong, Liaoning, Shanxi
Southeastern coastal areas	Fujian, Guangdong
Southwestern and southern regions	Yunnan, Guizhou, Sichuan, Guangxi, Xizang
Northeastern region	Liaoning, Jilin, Heilongjiang, eastern Inner Mongolia
Central region	Henan, Hubei, Jiangxi, Hunan, Anhui
Northwestern region	Shaanxi, Gansu, Ningxia, Qinghai, Xinjiang

basis delineate eight comprehensive economic zones. The proposal considered the different regional policies introduced by the central government for the four regions and incorporated the concept of regional economic zone. As a synthesis of the Seventh Five-Year Plan and the Ninth Five-Year Plan, the proposal added hardly anything new (Table 5.10).

Proposal of the National Development and Reform Commission for Eleventh Five-Year Plan: During the preliminary study of the Eleventh Five-Year Plan, the concept of national comprehensive economic zone was proposed again, which highlighted the role of the central government in macro management, including the overall distribution of population and cities, the development of major energy and agricultural production centers, the layout of international transportation and information hubs, and the conservation of ecological function zones of national importance. The proposal also paid attention to the dynamics of geopolitical and economic situations and considered the economic zones formed under the planned economy and its advantages. The proposal divided the country into nine comprehensive economic zones. Among them, Inner Mongolia, Henan, Jiangxi, and Hunan were broken up into different zones despite the provincial boundaries. The proposal was based on a new concept and put forward specific requirements related to the main functions and directions of the economic zones in terms of industrial development, resource protection, infrastructure development and urban layout. In addition, the proposal included the idea and requirement to prioritize the development of function zones (Table 5.11).

In addition, experts and scholars have put forward proposals based on their studies, including:

Liu Zaixing's Proposal: In 1985, Liu Zaixing proposed to divide the country into six comprehensive economic zones. He first developed the principles of economic zoning. The basic principle is to respect the internal similarity and regional differences. Other principles include combining the specialization formed in history and integrated development, developing economic centers and economic cores, creating

Table 5.10 The eight comprehensive economic zones

Economic zone	Scope
Northeastern comprehensive economic zone	Liaoning, Jilin, Heilongjiang
Northern coastal comprehensive economic zone	Beijing, Tianjin, Hebei, Shandong
Eastern coastal economic zone	Shanghai, Jiangsu, Zhejiang
Southern coastal economic zone	Fujian, Guangdong, Hainan
Mid-yellow river economic zone	Shaanxi, Shanxi, Henan, Inner Mongolia
Mid-yangtze river economic zone	Hubei, Hunan, Jiangxi, Anhui
Southwestern comprehensive economic zone	Yunnan, Guizhou, Sichuan, Chongqing, Guangxi
Northwestern comprehensive economic zone	Gansu, Qinghai, Ningxia, Xizang, Xinjiang

Table 5.11 National comprehensive economic zones under the ndrc 11th five-year plan

Name	Area (10,000 km2)	Gateway city	Geographic scope
Northeast economic zone	127	Shenyang, Dalian	Liaoning, Jilin, Heilongjiang, eastern Inner Mongolia (Hinggan League, Hulun Buir, Tongliao and Chifeng)
North China economic zone	91	Beijing, Tianjin	Beijing, Tianjin, Hebei, Shanxi, Shandong, central Inner Mongolia (Hohhot, Baotou, Ulanqab and Xilingol League), north-central Henan (Anyang, Hebi, Xinxiang, Jiaozuo, Shangqiu, Kaifeng, Zhengzhou, Luoyang, Sanmenxia and Puyang)
East China economic zone	45	Shanghai	Shanghai, Jiangsu, Zhejiang, Anhui, northern Jiangxi (Jingdezhen, Jiujiang, Shangrao, Nanchang, Yingtan, Fuzhou, Xinyu, Yichun and Pingxiang)
South China economic zone	75	Guangzhou, Hong Kong, Taipei	Guangdong, Guangxi, Hainan, Fujian, southern Jiangxi (Ganzhou and Ji'an), southern Hunan (Yongzhou, Chenzhou, Hengyang, Shaoyang), Hong Kong, Macao, Taiwan
Central China economic zone	43	Wuhan	Hubei, northern Hunan (Changsha, Zhuzhou, Xiangtan, Loudi, Huaihua, Xiangxi Autonomous Prefecture, Zhangjiajie, Changde, Yiyang and Yueyang), southern Henan (Xinyang, Zhumadian, Nanyang, Zhoukou, Xuchang and Luohe)
Southwest economic zone	113	Chengdu, Chongqing	Chongqing, Sichuan, Yunnan, Guizhou
Near northwest economic zone	110	Xi'an	Shaanxi, Gansu, Ningxia, western Inner Mongolia (Ordos, Alxa League, Bayan Nur and Wuhai)
Xinjiang economic zone	164	Urumqi	Xinjiang
Qinghai-xizang economic zone	192		Qinghai, Xizang

or improving economic links, avoiding the disruption of administrative divisions, adjusting the divisions when appropriate, avoiding the overlap of economic zones, and specifying the geographical boundaries. The proposal contained productive discussions about the linkage between comprehensive economic zones and the three regions, the role of economic hub cities and the function of the outlet to the sea, providing an important basis for later studies (Table 5.12).

Table 5.12 Six economic zones proposed by Liu Zaixing

Economic zone	Central city	Economic core	Scope
Northeastern zone	Shenyang, Dalian, Harbin	South-central Liaoning, Harbin, Changchun, and Jilin	Liaoning, Jilin, Heilongjiang, and eastern Inner Mongolia (Chifeng, Tongliao, Hulun Buir and Hinggan League)
North China zone (middle and lower reaches of the Yellow River)	Tianjin	Beijing, Tianjin, Tangshan, and areas along the Qingdao-Jinan Railway	Shanxi, Henan, Hebei, Shandong, Beijing, Tianjin, western and central Inner Mongolia
Central China zone (middle and lower reaches of the Yangtze River)	Shanghai, Wuhan	Yangtze River Delta, Wuhan-Dazhi region, Chagsha-Zhuzhou-Xiangtan region, Central Anhui	Hunan, Hubei, Jiangxi, Shanghai, Jiangsu, Zhejiang, Anhui
Southeastern zone	Guangzhou	Pearl River Delta	Fujian, Guangdong, Guangxi, Hainan, Hong Kong, Macao, and Taiwan
Southwestern zone	Chongqing	Chengdu, Chongqing, Kunming, Guiyang	Sichuan, Guizhou, Yunnan, and Xizang
Northwestern zone	Xi'an, Lanzhou	Central Shaanxi, Lanzhou	Shaanxi, Gansu, Ningxia, Qinghai, and Xinjiang

Chen Dongsheng's Proposal: Chen Dongsheng put forward a proposal in 1986 that envisaged six economic zones. Based on the horizontal economic linkage, he divided the country into four economic zones—the Northeastern zone, the Yellow River Basin, the Yangtze River Basin, and the Southern zone, with Xinjiang and Xizang treated as two special zones. Except for the northeastern region which remained unchanged, many provincial boundaries were broken—the traditional North China and Northwestern economic zones were replaced by the Yellow River Basin economic zone, while the East China, Central China, Southeastern and Southwestern economic zones were incorporated into the Yangtze River Basin and the Southern economic zones respectively. The proposal emphasized the role of major cities as economic centers in the development of infrastructure networks and the opening of land and seaports of entry. It also stressed the importance of ports of entry to economic zones and attached great importance to the role of outward-oriented economic links in economic zones (Table 5.13).

Yang Shuzhen's Proposal: In the Study on Economic Zones in China complied in the late 1980s, Yang Shuzhen put forward the theoretical basis for economic zones and proposed to divide the country into ten economic zones. The proposal took into consideration the regional disparities, population size, ethnic minorities,

Table 5.13 Six economic zones proposed by Chen Dongsheng

Economic zone	Central city	Port of entry	Scope
Northeastern economic zone	Shenyang	Dalian, Yingkou	Liaoning, Jilin, Heilongjiang, and eastern Inner Mongolia (Chifeng, Tongliao, Hulun Buir and Hinggan League)
Yellow river basin economic zone	Tianjin	Tianjin, Qingdao, Qinhuangdao, Rizhao, Lianyungang, Erenhot	Qinghai, Gansu, Ningxia, Shanxi, Shandong, Beijing, Tianjin, northern Henan, central Shaanxi, northern Shaanxi, western and central Inner Mongolia
Yangtze river basin economic zone	Shanghai	Shanghai, Bailun Port	Sichuan, Hubei, Anhui, Jiangsu, Zhejiang, Shanghai, southern Shaanxi, southern Henan, northern Jiangxi, central Jiangxi, northern Hunan, central Hunan, northern Guizhou
Southern economic zone	Guangzhou	Guangzhou, Zhanjiang, Xiamen	Yunnan, Guangxi, Guangdong, Fujian, southern Hunan, southern Jiangxi, southern Guizhou
Xinjiang			Xinjiang
Xizang			Xizang

and the presence of planned commodity economy as well as central cities, their economic gravity and the role of coastal port cities and inland border trade centers in regional economic collaboration. The proposal was comprehensive in the sense that it included multiple principles and objectives, but the emphasis on economic principles was insufficient (Table 5.14).

Gu Chaolin's Proposal: Applying the basic principles of graph theory and geographical agglomeration, Gu Chaolin proposed to divide the country into nine urban economic zones in the early 1990s. He viewed the city as a system including social, economic, technological, and educational links and a spatial structure organized by urban system and transport network. It was one of the first proposals that organized regional economic activities around cities, which was of great significance for understanding the central role of cities. However, the proposal of organizing urban economic zones based entirely on mathematical methods deviated from the complex socio-economic linkages, leading to inappropriate divisions (Table 5.15).

Hu Xuwei's Proposal: Hu Xuwei proposed to divide the country into six economic zones under five principles: smoothening the economic links between regions, facilitating regions and provinces to complement each other's strengths and collaborate with each other, implementing major infrastructure, resource development and environmental improvement projects across provinces, regions and municipalities, maximizing the influence of central cities and economic cores, and strengthening the planning guidance and macroeconomic regulation on cross-provincial and cross-regional economic cooperation organizations or corporate groups. The proposal adopted an

Table 5.14 Ten economic zones proposed by Yang Shuzhen

Economic zone	Central city	Scope	Economic zone	Central city	Scope
Northeastern Zone	Shanyang	Liaoning, Jilin, Heilongjiang	Southwestern Zone	Chongqing	Sichuan, Yunnan, Guizhou
North China Zone	Tianjin	Beijing, Tianjin, Hebei, Shanxi, Shandong	Northwestern Zone	Xi'an	Shaanxi, Gansu, Ningxia, Qinghai
East China Zone	Shanghai	Shanghai, Jiangsu, Zhejiang, Anhui, Jiangxi, Fujian, Taiwan	Inner Mongolia	Hohhot	Inner Mongolia Autonomous Region
Central China Zone	Wuhan	Henan, Hubei, Hunan	Xinjiang	Urumqi	Xinjiang Uygur Autonomous Region
South China Zone	Guangzhou	Guangdong, Guangxi, Hainan, Hong Kong, Macao	Xizang	Lhasa	Xizang Autonomous Region

Table 5.15 Nine urban economic zones proposed by Gu Chaolin

Economic zone	Central city	Scope
Shenyang (Northeast China)	Shenyang, Harbin	Liaoning, Jilin, Heilongjiang, and eastern Inner Mongolia (Chifeng, Tongliao, Hulun Buir and Hinggan League)
Beijing and Tianjin (North China)	Beijing, Tianjin	Beijing, Tianjin, Hebei, Shandong, Xilingol League in Inner Mongolia, northern Jiangsu, northern Anhui, northeastern Henan
Xi'an (Northwest China)	Xian, Lanzhou	Shanxi, Shaanxi, Ningxia, eastern Qinghai, eastern Gansu, western Inner Mongolia
Shanghai (East China)	Shanghai, Nanjing	Shanghai, Jiangxi, south-central Jiangsu, south-central Anhui, northeastern Zhejiang
Wuhan (Central China)	Wuhan	Hunan, Hubei, southern Henan, southern Shaanxi
Chongqing (Southwest China)	Chongqing	Sichuan, Yunnan, Guizhou, western Guangxi
Guangzhou (Southeast Coast)	Guangzhou	Fujian, Guangdong, Hainan, Taiwan, Hong Kong, Macao, eastern Guangxi
Urumqi (Far Northwest)	Urumqi	Xinjiang, Tsaidam Basin in Qinghai, Hexi Corridor in Gansu, Alxa League in Inner Mongolia
Lhasa (Qinghai-Xizang Plateau)	Lhasa	Xizang, Yushu Xizangan Autonomous Prefecture in Qinghai

approach that engaged multiple stakeholders, made a distinction between the important and the lesser one and explored the cross-over to delineating the economic zones, which was very feasible as the intricate economic links between provinces and regions were fully considered. However, the analysis of the relationship between provinces or regions and their respective core areas in the proposal was questionable (Table 5.16).

Zhou Yixing and Zhang Li's Proposal: Based on the levels of central cities, Zhou and Zhang delineated three primary urban economic zones with Beijing-Tianjin-Tangshan, the Yangtze River Delta and the Pearl River Delta their respective core areas and identified the core and outer core hinterlands of the three zones based on the analysis of the flows of foreign trade, railroad passenger and cargo, population movement and communications. In the end, they divided the country into three primary urban economic zones (Northern Zone, East-Central Zone, and Southern Zone) and 11 secondary zones. The proposal was of great importance for the role of the three metropolitan areas and the secondary economic zones under their influence. However, for lack of consistency with the administrative division, the proposal encountered obstacles at the operational level (Tables 5.17 and 5.18).

The previous analyses show that urbanization policy zoning should consider both the division of comprehensive economic zones and the characteristics of regional urbanization process. Therefore, urbanization policy zoning should follow the following principles: first, consider the consistency of regional geographical features (resources and environment) and maintain the continuity of regional history and culture; second, consider the closeness of social and economic ties within the region, the influence of the national industrial strategy, and the scope of influence of economic hub cities; third, consider the conditions for zoning, maintain the integrity of provincial and regional administrative units as much as possible, and increase the feasibility of the policy; fourth, consider the complete network structure of regional

Table 5.16 Six economic zones proposed by Hu Xuwei

Economic zone	Provinces, regions, and municipalities as primary members	Provinces, regions, and municipalities as secondary members
Northeastern zone	Liaoning, Jilin, Heilongjiang	Inner Mongolia (eastern)
Northern zone	Beijing, Tianjin, Hebei, Shanxi, Inner Mongolia, Shandong, Henan	Liaoning, Shaanxi, Jiangsu (northern), Anhui (northern)
Northwestern zone	Shaanxi, Gansu, Ningxia, Qinghai, Xinjiang	Xizang, Inner Mongolia (western)
East-central zone	Shanghai, Jiangsu, Zhejiang, Anhui, Jiangxi, Hubei, Hunan	Henan, Shaanxi (southern), Sichuan (southeastern)
Southwestern zone	Sichuan, Yunnan, Guizhou, Xizang	Guangxi, Hainan, Guangdong (western)
Southern zone	Guangdong, Fujian, Guangxi, Hainan	Hunan, Jiangxi, Hong Kong, Macao, Taiwan

Table 5.17 Three primary urban economic zones proposed by Zhou Yixing and Zhang Li

Economic zone	Central city	Core area	Core hinterland	Outer core hinterland	Competitive hinterland	Marginal hinterland
Northern zone	Beijing, Tianjin	Beijing, Tianjin, Tangshan	Beijing, Tianjin, Hebei, Shanxi, central Inner Mongolia	Liaoning, Jilin, Ningxia, Gansu, Qinghai, eastern Inner Mongolia (Chifeng, Tongliao, Hulun Buir and Hinggan League, Alxa League)	Shandong, Henan, Shaanxi	Xinjiang, Heilongjiang
East-central zone	Shanghai, Nanjing, Hangzhou	Yangtze River Delta	Shanghai, Jiangsu, Zhejiang, Anhui	Hubei	Shandong, Henan, southern Shaanxi, Jiangxi, Sichuan, Chongqing, Guizhou, Fujian	Hunan
Southern zone	Guangzhou, Shenzhen, Hong Kong, Macao	Pearl River Delta	Guangdong, Hunan, Guangxi	Hainan, Yunnan, Xizang	Jiangxi, Guizhou, Sichuan, Chongqing, Fujian	Hubei

infrastructure and the role of regional hub cities; and fifth, consider the effective management of regional resources, such as the conservation of cross-regional water resources and ecological resources. To sum up, urbanization policy zoning should be based on economic divisions, with one to two mega-cities as the core to organize regional economic activities and the gateway for the region to participate in global competition so that regional economic activities are organized, and the requirements of urbanization policy implemented. The zoning should consider administrative divisions, but not be completely bounded by them. This study proposes seven new urbanization policy zones based on the above principles, each of which has a mega-city as the core to organize economic activities and implement different urbanization policy requirements (Table 5.19).

Different urbanization policy requirements are put forward according to the zoning proposal:

Table 5.18 Secondary urban economic zones proposed by Zhou Yixing and Zhangli

Secondary zone	Central city	Core area	Hinterland
North China	Beijing, Tianjin	Beijing, Tianjin, Tangshan	Beijing, Tianjin, Hebei, Shanxi, central Inner Mongolia, northern Henan
East China	Shanghai, Nanjing, Hangzhou	Yangtze river delta	Shanghai, Jiangsu, Zhejiang, Anhui, northern Jiangxi
South China	Guangzhou, Hong Kong, Shenzhen, Macao	Pearl river delta	Guangdong, Hunan, Guangxi, Hainan, southern Jiangxi
Northeast China	Dalian, Shenyang	South-central liaoning	Liaoning, Jilin, Heilongjiang, and eastern Inner Mongolia (Chifeng, Tongliao, Hulun Buir and Hinggan League)
Southwest China	Chongqing, Chengdu	Sichuan basin	Chongqing, Sichuan, Yunnan, Guizhou
Northwest China	Xi'an, Lanzhou	Central Shaanxi, Lanzhou	Shaanxi, Qinghai, Ningxia, Gansu
Xinjiang	Urumqi	North slope of Tien Shan in Urumqi, Shihezi and Hami	Xinjiang
Xizang	Lhasa	Angqu, Zhaqu, Sequ and Lancang rivers	Xizang
Shandong	Qingdao, Jinan	Shandong Peninsula	Shandong
Fujian	Xiamen, Fuzhou	Southeastern Fujian	Fujian
Hubei	Wuhan	Wuhan Metropolitan Area	Hubei, southern Henan

(1) North China: Improve the quality of urbanization in Beijing-Tianjin-Hebei region, develop the Shandong Peninsula city cluster proactively, and foster the development of city and town clusters around Beijing and Tianjin in Hebei Province; improve the level of industrial modernization, develop high-tech industries and tertiary industries actively, and cultivate Tianjin Binhai New Area as a new economic growth pole; improve the collaboration between Beijing-Tianjin and Hebei, Shanxi and Inner Mongolia on energy utilization and environmental protection, and protect the environment of Inner Mongolia and northern Hebei in strict compliance with laws and regulations; strengthen the construction of cross-provincial transportation corridors, and coordinate the operation and management of the Port of Tianjin and Tangshan Port in Bohai Bay, Qinhuangdao Port and Huanghua Port in Hebei Province and Qingdao Port and Yantai Port in Shandong Province (Fig. 5.8).

Table 5.19 Proposal of urbanization policy zones in China

Policy zone	Central city	Core areas	Hinterland
North China	Beijing, Tianjin	Beijing, Tianjin, Tangshan	Beijing, Tianjin, Hebei, Shanxi, central Inner Mongolia, northern Henan, Shandong
East China	Shanghai, Nanjing, Hangzhou	Yangtze River Delta	Shanghai, Jiangsu, Zhejiang, Anhui, northern Jiangxi
South China	Guangzhou, Hong Kong, Shenzhen	Pearl River Delta	Guangdong, Guangxi, Hainan, Hunan, southern Jiangxi, Fujian
Northeast China	Shenyang, Dalian, Harbin	South-central Liaoning	Liaoning, Jilin, Heilongjiang, and eastern Inner Mongolia (Chifeng, Tongliao, Hulun Buir and Hinggan League)
Southwest China	Chongqing, Chengdu	Sichuan Basin	Chongqing, Sichuan, Yunnan, Guizhou, Xizang
Northwest China	Xi'an, Lanzhou	Central Shaanxi, Lanzhou	Shaanxi, Gansu, Qinghai, Ningxia, Xinjiang
Central China	Wuhan	Wuhan Metropolitan Area	Hubei, southern Henan

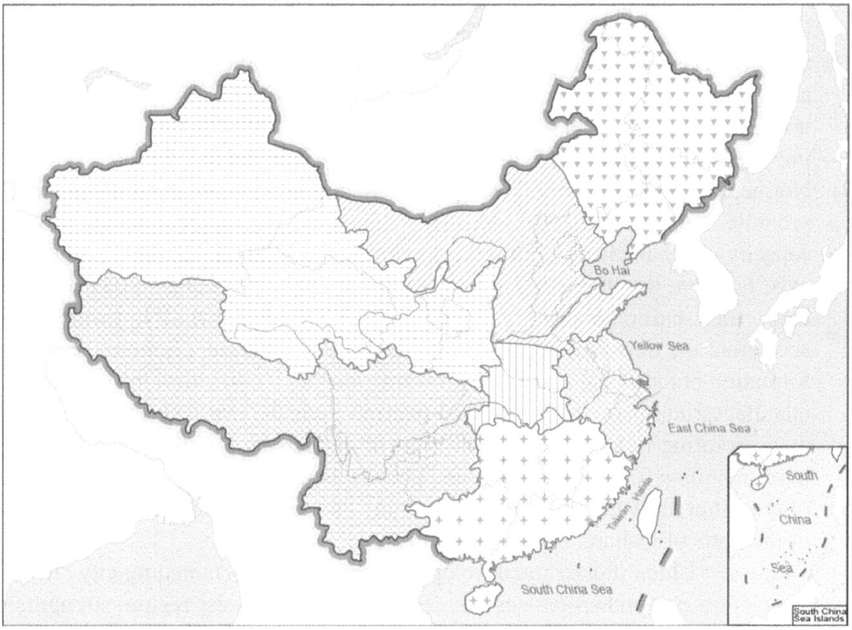

Fig. 5.8 Zoning diagram illustrating urbanization policy areas

(2) East China: Improve the quality of urbanization in the Yangtze River Delta, control the size of population, create an internationally competitive mega city regions with Shanghai as the core, expand the scale of core cities in the Yangtze River Delta, actively develop small and medium-sized cities, and create a networked city system; coordinate the development of industrial belts along the Yangtze River, curb high-energy-consuming industries, improve the technology of industrial clusters, and actively develop high-tech industries as well as cultural and creative industries, and improve the services of tertiary industries; strengthen the management of water resources, enhance the control of water pollution, and protect coastal beaches and wetlands; strengthen the development of intercity rail transportation, improve the overall infrastructure services of the region, and coordinate the operation and management of Shanghai International Shipping Center, Ningbo Zhoushan Port and Wenzhou Port in Zhejiang Province, Nantong Port and Nanjing Port in Jiangsu Province.
(3) South China: Improve the quality of urbanization in the Pearl River Delta, collaborate effectively with Hong Kong and Macao on regional functions and division of labor, maintain Hong Kong's status as the financial center in Asia, develop the city cluster on the west side of Taiwan Strait, and foster the development of Changsha-Zhuzhou-Xiangtan city cluster and Beibu Gulf city cluster; continue to enable the Pearl River Delta to play the role of an important global industrial base, improve the R&D capability and industrialization of high technology; strengthen the international cooperation capability of border cities such as Nanning, and improve collaboration with Southeast Asia in economy and trade; improve the level of regional transportation infrastructure services, and improve transportation integration in the Pearl River Delta.
(4) Northeast China: Actively develop the city cluster in south central Liaoning, promote the upgrading of industries, improve the comprehensive service capacity of urban clusters, promote the development of the Harbin-Daqing city axis, improve the international cooperation of Shenyang, Dalian, Harbin and other cities, and create an important economic and trade center in Northeast Asia; accelerate the transformation of traditional industrial base, promote the transformation of industrial structure in resource-depleted cities, develop equipment manufacturing, petrochemicals, and pharmaceuticals, and vigorously develop manufacturing-related service industries; strengthen the conservation of ecosystems including forests, wetlands, and phaeozem, and enhance the management of cross-border water resources; coordinate the development and management of the ports of Dalian, Yingkou and Jinzhou.
(5) Southwest China: Foster the development of Chengdu-Chongqing city cluster to create a comprehensive service center in the southwest region, strengthen and improve the functions of regional and provincial central cities based on the proper use of the scarce land resources, develop county-level central cities, promote international cooperation of border cities such as Kunming, and enhance economic and trade cooperation between China and ASEAN and the Greater Mekong region; develop manufacturing and resource-based processing

industries to avoid a reliance on industries with notable environmental disturbances to the maximum extent possible and minimize the impact on ecological environment from the source; strengthen the conservation of forests and water bodies to maintain biological diversity; enhance the development of transportation corridors between Southwest China, South China and East China to ensure smooth access to the sea and strengthen the development of cross-border road facilities to improve the economic development of border areas.

(6) Northwest China: Actively develop the city cluster in central Shaanxi, create a comprehensive service center in the northwest, focus on the development of cities and towns of moderate size considering the scare water resources in the northwest, gradually increase the concentration of population, areas unsuitable for urban development should encourage the implementation of environmental migration policies, improve the international cooperation of Urumqi to build a bridgehead for economic and trade cooperation between China and Central Asia; vigorously develop the circular economy, reduce the damage to the environment caused by resource-based industries, promote the development of high-tech industries, and determine the layout and scale of industrial development based on the carrying capacity of environment; reduce soil erosion caused by human activities, and build a system to control soil erosion and desertification with small basin as the unit; strengthen the development of corridors linking with the east, and give play to the role of the Eurasian Land Bridge in the areas of economy, trade and energy.

(7) Central China: Actively develop the Wuhan city cluster and the Central Plains city cluster, actively develop regional central cities and county-level central cities taking into account of the high population density and strong agricultural base to create a sound hierarchy of cities and towns; foster energy and manufacturing industries with comparative advantages, develop service industries that support agricultural industrialization and manufacturing, promote the division of labor among cities and towns, and encourage large and medium-sized cities to focus on the development of high-end industries with considerable technological sophistication and high added value, and phase out industries with low added value, high pollution and low resource efficiency while accepting the transfer of industries from the eastern region; strictly protect arable land resources, strike a balance between urban development and arable land protection, and enhance the prevention and control of pollution in the Huai River basin; strengthen regional infrastructure development, increase the access of the central region to the sea, and enhance the role of the central region in connecting the east to the west.

5.3.4.2 Establish a Unified and Efficient Public Service System and Promote Urban–rural Integration

The coordinated development of urban and rural areas, the biggest challenge facing China's urbanization at present, is both a necessary part of building the new socialist countryside and a prerequisite for improving the quality of urbanization. In the past

20 years, the income gap between urban and rural areas has increased from 2.6:1 in 1978 to 3.1:1 in 2003, and the gap of personal consumption expenditure between urban and rural residents has increased from 2.9:1 in 1978 to 3.5:1 in 2002. The gap in the quality of living environment is even greater. Most rural residents have no access to the basic public health and medical care. The rural population, which accounts for 70% of the total, only consumes 20% of the medical products and services. In terms of the distribution of educational resources, the average number of years of education received by the rural population aged 15 and above is less than seven. There are 85 million illiterate and semi-literate people nationwide, with more than 3/4 of them concentrated in the western rural areas, areas inhabited by ethnic minorities and national-level poverty-stricken counties. Therefore, the fundamental way to eliminate urban–rural disparity lies in providing equal access to public services, not by replicating the urban employment environment and living environment in rural areas, but by creating conditions for equal access to public services, to provide rural residents with an opportunity for their own development. The experience of Japan and Korea in rural development at the stage of rapid industrialization demonstrates that providing equal access to public services for rural residents in transportation, education and healthcare is of great importance. Given the realities in China, the following aspects should be improved in the development of urban and rural infrastructure:

(1) Transportation and Municipal Utilities: Extend the road network from urban areas to rural areas, give priority to the construction of road networks with major towns and villages and promote the development of inter-village public transit; develop the water supply facilities in towns and villages in an integrated manner and promote the use of clean energy such as biogas.
(2) Education: Give priority to popularizing and promoting nine-year compulsory education in rural areas and expand the coverage of high-quality educational resources; make extra effort to promote vocational education and adult education in areas with major labor inflows and outflows and further enhance exchanges and cooperation in education.
(3) Health and Medical Care: Establish the three major systems of medical services, preventive health care, and health supervision, establish and improve the rural cooperative medical care system, create a health security facility system covering both urban and rural areas, create a disease prevention and control system with quick response and strong supervision, and improve the response capability to sudden public health events.
(4) Culture: Promote the development of non-profit cultural programs, facilitate the prosperity of for-profit cultural industry, and build a relatively complete system of public cultural facilities of different levels to meet the needs of different groups in the society, especially the development of libraries and cultural centers, to meet the needs of districts and towns for cultural development.
(5) Social Welfare and Social Assistance: Gradually establish a social welfare management system and operating mechanism compatible with the socialist market economy and accelerate the development of social welfare and relief facilities such as pension service facilities, children's welfare service facilities,

social relief facilities, service facilities for people with disabilities and funeral service facilities.

At an appropriate time, the standards for public service facilities in rural areas will be introduced, and the land and funds for building the facilities will be made available in the process of planning and development, to effectively eliminate disparities between urban and rural areas.

5.3.5 Build a Dynamic and Diverse Urban Spatial Structure Based on the Uncertainty of Development

5.3.5.1 Develop a Globally Oriented and Dynamic Urban Spatial Structure

As analyzed above, the impact of economic globalization on the spatial structure of China is multifaceted, and it's precisely because of the impact of globalization that the urban spatial structure of China is changing more frequently and uncertainly than at any time in the past. Although the impact of capital on regional industrial space and the resulting urban space is relatively easy to judge in the near to medium term, with the flow of capital and a series of policy changes, there is a great deal of uncertainty in the spatial structure at the national level in the medium to long term. Therefore, the urban spatial structure in the medium and long term must provide a variety of options and respond to the multiple possibilities of development. Based on the characteristics of urban development in the era of globalization, Chinese scholars propose, firstly, to integrate China's urban system into the world city system as soon as possible, select one to three cities with international influence, and develop and cultivate them into world-class cities and international cities. Secondly, the overall development of several large and medium-sized cities is positively encouraged, especially for cities with more than 1 million population. Several large and medium-sized cities in the eastern, central, and western regions will be selected and cultivated into hubs that not only connect with the world city system, but also facilitate national and regional economic development, which will gradually become part of the world city network, rather than simply part of the regional city network in China. Thirdly, they attach importance to the development of contiguous urban areas, especially the Yangtze River Delta, the Pearl River Delta, the Beijing-Tianjin-Hebei Bohai Rim, and South-Central Liaoning.

The spatial structure of "one belt, seven axes and multiple centers" proposed in the outline of the new national urban system planning is honestly very innovative. It is an inheritance of the trend of urban spatial development in China in the past 20 years and meanwhile captures, to a considerable extent, new development trends. The concept of "multiple centers", the cultivation of new growth poles in the central and western regions and the focus on border cities are of considerable strategic importance. However, from the perspective of development, the macro-level spatial structure tends to be less stable, while the long-range plan tends to be more uncertain. Therefore, an open and resilient spatial structure is more valuable and conducive to the future urban spatial structure in China. Spatial policy guidelines for different regions are more meaningful than specific spatial patterns. The belt, axes and priority areas should be delineated based on the current needs of national development and be adjusted to suit the changing conditions at different points in time, while the development policies for "problematic areas" (dense urban agglomerations and "rust belt" areas) and the arrangement of spatial structures in smaller areas should be the focus of spatial planning. Therefore, it is more relevant to establish an open spatial structure that is resilient to multiple possibilities. Considering all the factors above, the urban spatial structure in China should be based on the following principles:

First, it must be linked up with the world city system, especially to integrate with the urban space of the Asia–Pacific region and create an open spatial structure. It starts with choosing mega city regions with international influence for global competition, and then develop them into international cities comparable with New York, London, Paris, and Tokyo. Considering the development level of cities in China, the most eligible ones are the Yangtze River Delta with Shanghai as the core and the Pearl River Delta with Hong Kong as the core, while the Beijing-Tianjin-Hebei Bohai Rim with Beijing and Tianjin as the core has the greatest potential. Then, it is necessary to enhance infrastructure development in the cities across the country, including transpiration and communication, to reinforce the close connections with the world, in particular the development of the cities that are identified as regional transport hubs by the urbanization policy zoning as previously stated (Beijing, Shanghai, Guangzhou, Wuhan, Shenyang, Xi'an, and Chengdu).

Second, carry out dynamic multi-scenario simulation according to the different stages of industrialization and urbanization in China and their development trends. Generally speaking, China is in the middle stage of the industrialization process, characterized by the extension of manufacturing industries cross the value chain both upstream and downstream, and typical examples include the continuous expansion of petrochemicals industry, increasing concentration of urban space in the coastal areas and slower development in the central and western regions; with the upgrading of industrial structure and the development of tertiary industries such as service industry, the urban space and functions in the coastal areas, especially the megacities, tend to decentralize, while the development of cities and towns in the central and western regions tend to centralize; and with further development, the overall functional structure and spatial structure of the cities and towns across the country will be increasingly networked. In view of the development trend, the spatial structure plan should be adjusted to suit the changing situations at different points in time.

Third, it is necessary to consider the space (cities and regions) needed for pursuing international cooperation in the fields of foreign trade and energy utilization. In addition, the infrastructure for transnational cooperation should also be included in the overall spatial planning, in particular resource, energy, and transport infrastructure.

5.3.5.2 Build the Main Body of Urban Spatial Structure Centering Around City Clusters

City cluster is a manifestation of regional urbanization and an advanced form of urbanization. As the urbanization process deepens in China, several economically developed and closely connected dense city regions have appeared in the coastal areas. In government work reports and academic studies, city cluster has been included as an important concept. In the Outline of the 11th Five-Year Plan for National Economic and Social Development, city cluster is considered as the main form of urbanization, and it is proposed that new urban clusters should be cultivated under appropriate conditions as part of the proper spatial layout of urbanization[3].

There is no single agreed-upon definition of city cluster, but the first one originated from the theory of "megalopolis" proposed by French geographer J. Gottmann (1957) when he studied the northeast coast of the United States. He believed that the leading form of economic space in the northeast coast of the United States was the giant chunk consisted of close links formed by economic activities in densely populated metropolitan areas, which is an urbanized areas linked up by several metropolises and an urban zone with a certain population density. He predicted that "megalopolis" was the most advanced stage of human settlement and city cluster. He identified the northeast coast of the United States, the east coast of Japan, the southeast and central parts of England, the metropolitan region of northwestern Europe, the Great Lakes metropolitan region of the United States and Canada and the Yangtze River Delta of China as the six largest urban regions in the world. In the subsequent studies, several scholars introduced the concepts of "metropolitan region", "mega-city region", "urban agglomeration" and "metropolitan area."

Although there is no single agreed-upon definition of city cluster, it basically refers to the concentration of numerous cities and towns within a particular geographic area. Located near each other, the cities and towns of varying functions and sizes center around one or several central cities as the core of regional economy and constitute a close-knit network of urban areas with the efficient comprehensive transportation network and sophisticated information network as the backbone [12–16]. At an international conference on megacity held in Tokyo in 1992, Wu Liangyong presented his paper which pointed out that "city regions" were emerging in megacities such as Shanghai and forming a system of megacity regions. In the book Introduction to Science of Human Settlements, he pointed out that "the megacity is not simply a network system with a particular megacity as the core or surrounded by several satellite cities, but instead the integrated development of the entire region that combines points, lines, and textures and takes the form of multicore city cluster [17]. So technically speaking, there are only three well-developed urban clusters in China—the Yangtze River Delta, the Pearl River Delta, and the Beijing-Tianjin-Hebei region.

---

[3] In the *Outline of the 11*th *Five-Year Plan for National Economic and Social Development*, the concept of "city cluster" was explicitly proposed as part of the proper spatial distribution of urbanization, which was not found in any national economic and social development plans in the past.

However, with the progress of urbanization, city cluster, a unique urban spatial form, is bound to play an increasingly important role in urbanization in future given its huge economic impact and remarkable space efficiency. In the section about urbanization policy zoning, the policy zones are delineated based on regional central cities to highlight the importance of coordinated development between regions. Similarly, to build the national urban spatial structure based on urban clusters can also effectively increase space efficiency and promote the development of growth poles in the regions, thus driving regional economic growth. However, the determination of a city cluster should be based on the following criteria: first, there is a relatively complete city system within the city cluster, where the cities and towns play their respective roles, and collaborate closely with each other to deepen and develop the division of labor and collaboration between core cities, and the cities within the cluster are able to innovate and evolve constantly; second, there are one to two core cities with well-developed economy, sound functions, at least 2 million population and comprehensive management functions that are qualified as the gateway city and able to lead the surrounding areas to become the growth pole in regional economic development; third, the population and economic size of the cities and regions in the city cluster account for higher proportions in the regional total, close to the status of primate city, the population density is mostly between 400 and 1500 people per square kilometer or more and the industries in the cities, towns and rural areas are closely connected and complement each other; fourth, there is a well-developed infrastructure network within the city cluster, the central city serves as the comprehensive transport hub and gateway city in the region and the travel time between major cities is within three hours; and fifth, the city cluster has a regional collaboration mechanism that is able to coordinate the development of major infrastructure projects through a series of regional development policies and plans.

According to the above criteria and in light of future levels of economic development, there are 16 existing and potential urban clusters: in the eastern region, there are Beijing-Tianjin-Hebei, Yangtze River Delta, Pearl River Delta, Shandong Peninsula, and the West Side of the Strait; in the central region, there are urban clusters centered on Zhengzhou, Wuhan, Taiyuan, and Changsha-Zhuzhou-Xiangtan; in the western region, there are clusters centered on Xi'an, Lanzhou, Chengdu-Chongqing, and Kunming; and in the northeast, there are clusters in south-central Liaoning, Changchun and Harbin. These urban agglomerations are largely located in the east of the Aihui-Tengchong line. With only 12.71% of national land area, they are home to 36.2% of the total urban nonagricultural population in the country, the main source of local government revenue and the important import and export base in the country, contributing 76.3% of the total economy and 65.9% of the total retail sales of consumer goods nationwide. In conclusion, to build the national urban spatial structure with urban clusters as the mainstay is in line with the development trend of urbanization and conducive to the intensive use of space, which would play an effective role in promoting coordinated regional development and improving the overall competitiveness of the country (Table 5.20).

Table 5.20 Major economic indicators of urban clusters in 2003 as a percentage of the national total

	Area (%)	Non-agricultural population (%)*	GDP (%)	Retail sales (%)
Beijing-Tianjin-Hebei	0.95	4.27	8.42	8.43
Yangtze River Delta	1.20	7.09	19.87	15.28
Pearl River Delta	0.77	3.25	9.96	9.04
South-Central Liaoning	1.17	3.18	5.04	4.57
Shandong Peninsula	1.16	4.2	9.26	6.69
Chengdu-Chongqing	2.14	3.57	5.67	5.35
Xi'an	0.58	1.21	1.49	1.43
Wuhan	0.53	1.75	2.53	2.90
Zhengzhou	0.51	1.44	2.29	1.91
West Side of the Strait	0.57	1.38	3.94	3.24
Changsha-Zhuzhou-Xiangtan	0.42	0.89	1.55	1.69
Harbin	0.8	1.1	2.17	1.60
Changchun	0.5	1	1.65	1.50
Taiyuan	0.29	0.69	0.72	0.65
Lanzhou	0.44	0.61	0.59	0.66
Kunming	0.68	0.59	1.18	0.94
Total	12.7	36.2	76.3	65.9

Source Compiled from the research reports of "National Urban System Planning".
*The "non-agricultural population" refers to the non-agricultural population in the urban areas

5.3.5.3 Establish the Support System and Service System for Urban Development with Transportation as the Core

The development level of urban space depends largely on the supply of infrastructure. In terms of national spatial development, whether less developed regions can prevent themselves from being marginalized hinges on the accessibility of their locations and the degree of connection between its regional economy and that of the developed regions and even the world economy brought about by this accessibility. The EU Spatial Development Perspective emphasizes that the status of regional centers depends on if they are regional and even intercontinental transport hubs, while the prevention of marginalization of underdeveloped regions depends primarily on the supply of transport and information infrastructure.

Therefore, a hierarchical regional transportation network covering all major spaces in the human settlements is a crucial part of a sound national spatial structure. The purpose of developing transport hubs in North China, East China, South China, Central China, Northeast China, Northwest China and Southwest China with Beijing, Shanghai, Guangzhou, Wuhan, Shenyang, Xi'an and Chengdu as the center, respectively, mentioned earlier is to enable the cities to play their roles in regional transport organization and management and make use of their transport facilities to facilitate interactions with the outside as the gateways. In conclusion, to create a safe, efficient, and green transportation system compatible with the environmental and resource conditions of China and promote integrated multi-modal transport pivoting on transportation hub cities and based on better services is an important measure to promote the coordinated development of the eastern, central, and western regions. At the same time, how the government, which corrects market failures under the market economy, provide equal access to public services is the key for different regions to become civilized society. For underdeveloped regions, the improvement of transport infrastructure plays an irreplaceable role in regional economic development. In terms of planning, intergovernmental fiscal transfers should be increased to support the development of transport in underdeveloped regions and rural areas, enhance the comprehensive transport corridors and accelerate the development of secondary trunk roads and tributary networks, to create a well-structured hierarchical network that promotes the common development of transport in urban and rural areas and provide strong support for building a new socialist countryside.

5.3.6 Promote the Transformation of Urban Economic Growth Guided by New Industrialization

5.3.6.1 Develop the Concept of New Industrialization in Response to the Pressure and Challenge from Development

Industrialization is a development stage a country cannot skip in the process of modernization, and a specific historical stage and process of transition from an agricultural-based economy to an industrial-based economy. During the industrialization period, the proportion of industrial added value in the national economy rises continuously which drives the urbanization process. During industrialization over the past 20 years, industrialization has effectively encouraged urbanization, and likewise urbanization has greatly boosted industrialization. However, the perception of industrialization, both in the coastal areas and the inland, is still confined to the mode of traditional industrialization. The evidence can be found in the fact that numerous coastal and riverside cities proposed to take steel and petrochemical industries as the dominant industries in the future, and the central and western regions proposed to accept the energy-intensive, water-consuming, and polluting industries shifted from the east. In fact, under the influence of the new technological revolution, the concept of economic development centering around traditional industries is facing unprecedented challenges.

(1) Resource intensity is under challenge. Traditional industrialization is characterized by the massive consumption of resources including energy, raw materials, natural resources and tangible capital, and the development of heavy industries is a goal at a certain stage of industrialization. Modern high-tech industries

and information industries, however, are knowledge—and technology-intensive with much lower dependence on energy, which fundamentally reshapes the concept of raw material consumption and makes a significant change in the phenomenon of resource constraints and capital constraints in traditional industrialization.

(2) Traditional demand is under challenge. The goal of traditional industrialization is to meet people's material needs, and its growth is constrained by both supply and demand. With the considerable abundance of material goods, the growth rate of industrialization will continue to slow down when it is difficult to find new growth in the field of material products. This is the main reason why the economic growth rate of developed countries is generally lower than that of most developing countries before embracing the new economy. In contrast, the development of modern information and high-tech industries creates a large amount of new demand in the field of non-material products, and the new demand drives economic development for quite a long period of time, of which the evidence can be found in the rapid growth of high-tech enterprises in the U.S. and the resulting growth of the U.S. economy in the last decade.

(3) The traditional international division of labor is under challenge. In the last two decades of the twentieth century, traditional manufacturing industries, especially low value-added, resource-intensive, and highly polluting industries, continue to shift to developing countries. As a result, these countries face resource and environmental crises while their economies grow, and their status in the international division of labor has not fundamentally improved, which is particularly prominent in China. But this situation cannot sustain in developing countries which is why the Chinese government introduced the development strategy to build an "innovative country".

(4) The challenge from informatization. Informatization is another huge revolution in the economic history of the world, which in essence is the process of replacing part of the mental labor with computers. In the information age, knowledge (human capital) has displaced physical capital to assume a dominant role in economic activities and production processes, resulting in a change of the times. In contrast, the economy dominated by factors of industrial production is at a relative disadvantage.

In short, the path to new industrialization is a strategic choice made by China after weighing the pros and cons of traditional industrialization in the past 20 years and in response to the trend of industrial development in the world. The characteristics of new industrialization include: to drive industrialization by informatization and vice versa; to improve the quality and efficiency of economic growth by advancing scientific and technological progress and improving the quality of the labor force; to take the path of sustainable development by attaching importance to the balance between economic development and ecological conservation; to give full play to the advantages of human resources by properly dealing with the relationship between improving productivity and expanding employment.

5.3.6.2 Accelerate the Adjustment of Industrial Structure and Update the Urban and Regional Industrial Structures

To achieve the development goals of new industrialization, we should first deepen the industrialization process nationwide and promote the coordinated development of the eastern, the central, and the western regions. Second, we should rationalize and modernize industrial and economic structures and develop a more sophisticated industrial structure. Third, we should accelerate the industrialization of agriculture and properly deal with the relationship between industry and agriculture. Fourth, we should transform the mode of industrial and economic growth, pursuing intensive economic growth and improving economic efficiency through technological progress. Fifth, in terms of the implementation mechanism, we should give a bigger role to market mechanism, to effectively transform the functions of government. In the case of China, the main elements of new industrialization include: to use the latest technological achievements amid the trend of informatization worldwide to drive industrialization, complete the industrialization process within a shorter period of time, and enable informatization and industrialization to reinforce each other; to promote industrialization in an energy-efficient and environment-friendly manner with the aim of supporting sustainable development, and create an energy-efficient and environment-friendly industrial system that is oriented to sustainable development and fits the realities in China; to pursue innovations in key technologies and thus provide technical support for the leaps and bounds of industrialization and promote the translation of scientific and technological innovation; to shift from labor-intensive industries toward capital-intensive and technology-intensive industries and meanwhile strike a balance between capital-intensive and labor-intensive industries when it comes to the technological roadmap for industrialization and the choice of the leading industry, so as to make full use of the labor resources.

Overall, China is at the beginning of the middle stage of industrialization with great disparities in the industrial structures among different regions. In the central and western regions, agriculture, and light industry, which had played a leading role in the early stage of industrialization, need to be further developed. In the eastern and central region, heavy chemical industry, which had played a leading role in the middle stage of industrialization, will be enhanced. In the eastern coastal areas, emerging industries, which have been playing a significant role in the late stage of industrialization, will thrive. To sum up, the evolution of industries will show the following features:

(1) Industrial Clustering. With the development of the theory of industrial organization, the industrial cluster theory has linked the development of enterprises with the competitiveness of the country. With the advancement of marketization and globalization, the region-based industrial space has continued to develop.
(2) Industrial Integration. Industrial integration refers to the evolutionary process of different sectors or different industries in the same sector interpenetrating and intersecting with each other to gradually form new industries. In essence,

industrial integration is an innovation, which has weakened the substitution relationship between new industries and previous industries.

(3) Industrial Softening. Industrial softening has two meanings: one is the softening of the whole industrial structure, that is, the rapid development of the tertiary industry and the proportion of primary and secondary industries decreased; the other is the softening of each industry, that is, an industry becomes knowledge driven.

(4) Industrialization of high-tech products and integration of high-tech with industries. Industrialization of high-tech products refers to the transformation of high-tech research and development results into real productivity and volume production through technology development, product development, production capacity development and market development, thus fostering the high-tech industry that plays a leading role in maintaining strong and stable economic growth in China. At the same time, high technology will also continue to penetrate and spread to traditional industries, transform traditional production processes, increase technological sophistication, advance technological and product upgrades in traditional industries, and embed high technology into traditional industries.

In short, the development of cities and regional industries is facing new challenges. As to space, the 11th Five-Year Plan and the new round of national urban system planning both put forward the concept of 5 major metropolitan economic zones and 10 major population-industrial agglomerations and made specific predictions of the spatial development of each major category of industries. Regardless of their accuracy and feasibility, it is a certain assertion that the spatial structure of industries nationwide will undergo a major change in the next 20 to 30 years. Whether cities in different regions can achieve sustainable industrial development in this transformation process ultimately depends on whether they can truly follow the path of new industrialization by considering the objective conditions and development reality of the regions.

5.3.7 Establish a Planning System Based on Efficient Spatial Management and Integration of the Three Plans into One

5.3.7.1 To Establish a Regulatory System for Coordinated Socioeconomic Development with Spatial Planning as the Mainstay

In a market economy, economic activities are mainly regulated by the market. Since China established the socialist market economy in 1992, especially after her accession to the World Trade Organization in 2003, the marketization degree of China has reached 73.8%[4] with the improvement of the market system and exceeded the supercritical level of marketization (60%), becoming a market economy in development. The economic freedom of China has surpassed that of many transitional or developing countries recognized as market economies by Europe and the United States. Industrial development is basically regulated by the market, and even the formulation of major national industrial policies primarily considers the needs of the international and domestic markets. In countries with relatively complete market economic systems, such as Japan, South Korea and EU members, the means they used to regulate economic development are spatial policies and spatial planning, in addition to interest rate and taxation policies. The fundamental reason is that one of the key factors of economic development is location, and the influence of spatial planning on location is crucial. In fact, the economic system of western developed countries has already transitioned from a completely free economy to a controlled and guided market economy.

At the same time, spatial planning, which covers several objectives such as social development and ecological and environmental protection, has also become a comprehensive development plan for these countries. Therefore, under the socialist market economy, the means of regulation and control of China's comprehensive and coordinated socioeconomic development will certainly shift from the traditional "national economic and social development plan" with notable vestiges of the planned economy to the "national spatial planning" centering around the sound allocation of spatial resources. In fact, the National Development and Reform Commission has greatly changed its approach in the preliminary study and preparation of the 11th Five-Year Plan, which has increased the proportion of spatial planning and put forward more development guidelines from the perspective of the protection and rational use of national spatial resources. It is also believed that, during the transitional period, spatial planning and medium—and long-term national economic planning can jointly act as the two "wheels" of coordinated socio-economic development [18], with the former focusing on the macroeconomics of economic growth rate, fiscal balance, money supply and other aggregate balance, and the latter on the proper allocation of resources and productivity, namely "the book to account the national economy and the map to navigate the spatial development". However, in view of the development trend, spatial planning will inevitably replace the traditional national economic and social development plan.

5.3.7.2 To Establish a Unified and Efficient Spatial Plan Formulation and Management System Integrating the Three Plans into One

At present, the spatial development plans at the national level mainly include the "Five-Year Plan for National Economic and Social Development" formulated by the National Development and Reform Commission, the "National Urban System Plan"

[4] Quoted from *Report on the Development of China's Market Economy 2005*, Beijing: China Commerce and Trade Press.

prepared by the former Ministry of Construction and the "National Overall Land Use Planning" prepared by the then Ministry of Land and Resources as well as various specialized plans prepared by numerous ministries and commissions including transportation, railway, civil aviation, information, and environmental protection. As different plans are the charge of different departments, and the preparation time and duration of them are different, there are many conflicts between the plans. In particular, the specialized plans that underpin the socioeconomic development such as transportation and railway are not complete and systematic since there is no national urban development plan on which they are based. The 11th Five-Year Plan prepared under the auspices of the National Development and Reform Commission attaches importance to spatial planning, but there is still no systematic study of urban spatial development. Similar problems exist in the "National Overall Land Use Planning" prepared by the then Ministry of Land and Resources.

The above-mentioned intertwined spatial plans all center around the rational use and management of national spatial resources, and the focus of these plans is on the spatial organization with the urban system as the core. Therefore, at the institutional level, it is important to thoroughly adjust the current planning agencies under different ministries and commissions (especially the National Development and Reform Commission, the former Ministry of Construction and the then Ministry of Land and Resources) and establish a national spatial planning management agency led by a comprehensive department and a planning system that "integrates the three plans" to ensure the rational use of spatial resources in China.

The technical content of the unified spatial planning should mainly include the following: the prediction of population and economic growth and the analysis of the carrying capacity of resources and environment; the framework of regional spatial layout; the functional positioning of the urban system and major cities; the division of various functional zones such as industrial agglomerations and their positionings; infrastructure and other major projects for coordinated development; and regional coordination mechanism and other measures to guarantee the implementation of the plan. Objectively speaking, the establishment of a complete and unified spatial planning system is an important measure and means to guarantee the scientific and rational use of spatial resources.

5.3.7.3 To Establish Planning Implementation Measures and Instruments Based on Economic Policies Including Finance and Taxation

Basically, the experience of other countries in spatial plan implementation indicates that fiscal and taxation instruments are the most important measures. Since the 1950s, the European Union (formerly the European Community) has used "structural funds" as the main instrument to support underdeveloped regions and revitalize declining regions. The policy was further improved after the 1980s by verifying applicant eligibility for structural funds based on regional rather than national GDP per capita, and by actively intervening in regional disparities at a relatively micro level. For example, as the GDP per capita of Scotland and Wales of the United Kingdom is lower than the EU average, they are eligible for funds, while the GDP per capita of the UK is higher than the EU average (Figs. 5.9 and 5.10).

Therefore, the coordinated spatial development of China must highlight the application of economic instruments and focus on the following four areas: first, establish a "national regional development fund" to support the development of less developed regions and the revitalization of declining regions; second, use national fiscal policy such as mobile payment to promote the protection and rational use of resources in sensitive regions; third, use monetary policies to give full play to the role of policy banks and regional banks to promote local economic development; fourth, develop regional labor and employment policies to promote the rational flow and optimal allocation of labor resources between regions while preventing the loss of knowledge-based labor force in less developed regions and promoting the coordinated development of regions. In conclusion, the effective implementation of economic instruments, such as finance and taxation, is the guarantee that the spatial plan can be carried out (Table 5.21).

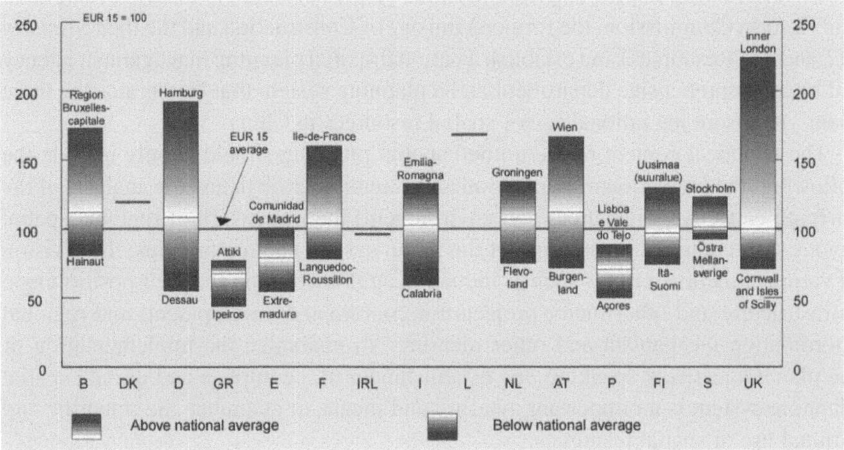

Fig. 5.9 Economic development levels of the EU-15. *Source* ESDP, European Commission 1999

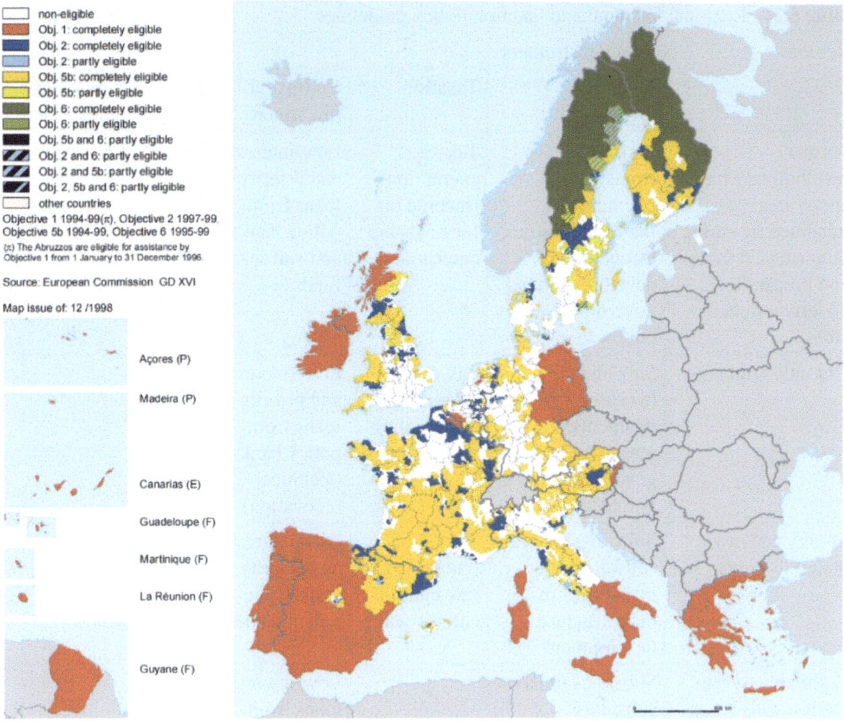

Fig. 5.10 Structural funds policy area. *Source* ESDP, European Commission 1999

5.3.8 *Establish the Think Tank for Urban Spatial Planning and Facilitate Sound Decision-Making on Urban Planning*

5.3.8.1 Establish the Special Committee on Spatial Planning of the National People's Congress to Regulate Planning Legislation

The previous analyses indicate that all departments under the State Council are preparing various comprehensive and special plans to put forward their own ideas on the spatial development of the country. In view of the current situation, it is difficult to make fundamental improvements in the short term, although since 2001, the Department of Development Planning of the National Planning Commission (now the National Development and Reform Commission), together with local planning commissions and scientific research institutions, has carried out research on the planning system, and published a book entitled "Theoretical Discussions on Planning System Reform" in February 2003, which for the first time explicitly included "planning" in the national economic plans prepared by planning departments and urban

Table 5.21 Regional financial and taxation policy guidelines

Region	Policy instrument				
	Finance	Taxation	Credit and investment	Price	Labor and employment
Former revolutionary base areas, areas inhabited by ethnic minorities, remote and border areas, poverty-stricken areas	Subsidies (transfer payments), interest subsidies, infrastructure development and government purchase	Tax exemptions (income tax for township enterprises)	Low-interest and priority loans from policy banks and various funds		Encourage labor export
Old industrial bases	Subsidies (transfer payments), interest subsidies, government purchase	Tax relief, accelerated depreciation	Low-interest and priority loans from policy banks (technical renovation)		
Energy and raw materials bases	Subsidies, interest subsidies, investment in infrastructure development	Tax relief (energy and raw material enterprises)	Low-interest and priority loans from policy banks		
Commodity grain (cotton, Oil) bases	Subsidies, interest subsidies		Low-interest loans from policy banks		
Overly densely developed regions	Relocation subsidies, interest subsidies		Relocation priority loans	Raise land prices	Restrict labor inflows
Regions with prominent environmental problems	Environmental infrastructure development and investment, relocation subsidies, interest subsidies		Relocation priority loans	Raise land and water prices	

Source Adjusted from the table in Theory and Methodology of National Territory Planning p421

spatial plans prepared by development departments, and put forward the idea of integrating various types of plans. According to the book, the basic principle of planning should be transformed from the three major balances of "investment, finance, and credit" in the planned economy to the new three major balances of "supply, demand and space" in the market economy. The main framework is to position the outline of the five-year national economic and social development plan as the "national master plan", incorporate the balance of spatial structure and the coordination of spatial development into the plan, and enhance the spatial guidance and constraint function of the "national master plan". The research proposes to "define spatial types and functions, and specify development directions and principles for densely developed urban or industrial areas, ecological and environmental protection areas, resource protection or development areas, natural disaster prevention and control areas, tourism and leisure areas, and agricultural areas; set out the goals of improving the urban system and adjusting the urban layout, the pattern of expansion of major metropolitan areas, and the direction and scale of population movement; set forth the improvement of land use structure, the number of land occupation, and regional configurations; to put forward the framework of the national infrastructure network system including transportation, electricity and water resources." It suffices to say that the proposal includes all the spatial elements, especially the elements of the spatial system of cities and towns. The most representative view is that "...... planning refers to the plan prepared by the government, which includes the strategy of national economic and social development in time and space and its specific arrangements." In fact, this idea has been fully reflected in the 11th Five-Year Plan.

However, it should be noted that the expansion of the national economic and social development plan into a master plan that integrates socio-economic and spatial development is not well-founded from a jurisprudential perspective, as there is no National Economic and Social Development Planning Law. At the same time, the planning system proposed from a departmental perspective to disrupt the existing system is inevitably distinctly departmental, even though the NDRC has the capacity to unify the plans given its status and authority. However, the implementation of the planning functions defined by laws and regulations including the Urban Planning Law, and the Land Administration Law is an issue that needs to be carefully sorted out. The approach of the central government to integrating various spatial resources and coordinating national socio-economic development with macro regulation means suggests that it is difficult for any single administrative department to propose a neutral and appropriate framework for planning system reform. Therefore, it is an impartial and fair solution to establish a "special committee on spatial planning" in the National People's Congress to sort out the plans prepared by various departments under the State Council and integrate them from the perspective of legislation. The composition of the committee should mainly include various experts in spatial planning research, who make suggestions on planning legislation from the perspective of a third party. In fact, the planning committees established at different levels in Japan, Singapore, and the United States, which are composed of planning professionals, parliamentarians, and relevant government officials, are good examples of giving advice on planning legislation and implementation. Some cities in China, such as Shenzhen, have set up urban planning committees in recent years, and they are also successful in making decisions from a professional and technical point of view on urban development and construction. The famous "Barlow Report"[5] in the history

[5] The *1940 Barrow Report* is a famous event in the history of urban planning in Britain. The report pointed out that the growing concentration of industry and population in London was due to the attraction caused by industry, and thus recommended the evacuation of industry and population in central London.

of urban planning in Britain is an excellent example of a professional committee composed of experts who play a significant role in national planning decisions.

5.3.8.2 Improve the National and Provincial Planning Inspection Systems to Facilitate Sound Decision-Making on Regional Space

Despite the remarkable achievements in urban development and construction in China so far, the problem in the supervision of implementation is still a common concern in both academic and political circles. Scholars believe that there are three things that are absent: the absence of guidance, coordination, and supervision from the provincial and municipal (prefecture) governments on urban and rural planning at lower levels, the absence of effective ex ante and ex post supervision during the implementation of urban and rural development plans, and the absence of restriction on "the will of the leader" and "confused instructions" [2]. The common practice in western countries is to increase the number of technocrats in government administration to facilitate sound decision-making on administration and make the implementation of plans more effective, and to provide ex ante advice and ex post supervision for decision-making from a professional point of view. With respect to theory, F. J. Goodnow argued from the perspective of political science that: "In fact, a large part of administration is unrelated to politics; therefore, it should be largely, if not entirely, freed from the control of political groups. The reason why administration is unrelated to politics is that it includes semi-scientific, quasi-judicial, and quasi-commercial activities—activities that have little, if any, influence on the expression of the true will of the state. To best serve this branch of the executive function, it is necessary to create a set of political institutions that are completely free from political influence[6]." Establishing a planning inspection system with professionals and technicians as the backbone is an effective way to facilitate sound and practicable decision-making on urban spatial planning. It is argued that having a professional or technician appointed by a higher level of government as a planning inspector can completely eliminate the interference from the head of local government in the recommendation and appointment of urban planners as well as in the superior-subordinate relationship and allow the technical decision-making and general administrative decision-making to give full play to their ability to create checks and balances [2].

Therefore, considering China's political system and national conditions, the establishment of a top-down planning inspection system with professionals and technicians as the backbone is an important way to guarantee sound spatial planning decisions and implementation. The specific duties of the national planning inspectors mainly include: first, to supervise the implementation of the national urban spatial planning, stop the behaviors in clear violation of the national planning principles and specific plans and request the higher authority to mete out punishment; second, to propose and veto the approval, site selection, and development of major

[6] Cited in *Urban Planning Reform in China's Urbanization Process* by Qiu Baoxing. Shanghai: Tongji University Press, 2005. 246.

cross-regional projects, especially infrastructure projects; third, to organize cross-provincial administrative authorities in charge of planning to make coordinated efforts on resources, environment and infrastructure development; and fourth, to make suggestions on the improvement of national historical and cultural cities, towns, and villages and national parks in the process of planning and construction.

5.3.8.3 Establish the Position of Chief Planner at Provincial and Municipal Levels to Facilitate and Supervise the Implementation of National Spatial Planning

One of the undertakings that numerous experts and scholars have been committed to in recent years is to establish the position of Chief Planner in cities and facilitate sound decision-making [2, 17]. The position has already been set up in Russia, the United States, Germany, and France. In terms of duties, the Chief Planner is a consultant to the decision-making on urban development and represents professional authority; and in terms of administrative structure, the Chief Planner is considered as a civil servant, or technocrat, but the tenure is not affected by the "cabinet reshuffle". Because of the above features, the Chief Planner can provide expert advice on the preparation, review, implementation, and revision of plans, to maximize the continuity of existing planning decisions and ensure the coherence, consistency, and continuity of city government's decisions, breaking out of the "tenure dilemma".

In fact, on this basis, to establish the position of Chief Planner in provinces (autonomous regions) is also an important means to guarantee sound decision-making of provincial governments on planning. With the rapid development of cities and towns in China, the decision-making of provincial governments on spatial planning is equally mindless. For instance, in recent years, some coastal provinces have implemented massive land reclamation projects in violation of the original plans in order to increase the supply of land for urban development, developed petrochemical industries without regard to the carrying capacity of natural resources along the coasts and rivers, initiated large-scale water infrastructure and water diversion projects for local economic interest without regard to natural conditions and promoted the IPO of scenic spots for short-term economic benefits. These anomalies are driven by local interest with the absence of expert knowledge. Therefore, to establish the position of Chief Planner in provinces (autonomous regions) can compensate for the fault in decision-making to some extent and facilitate sound decision-making of provincial authorities on urban development.

About technical support for the implementation of the national spatial planning, as the national urban spatial plan is a medium—and long-term development plan with duration from 15 to 20 years, the principles and main contents specified in the plan should be upheld and pursued in the long term. The provincial government is required to show the long-term commitment to the principles and the implementation of the technical part of the plan, and the Chief Planner in the province (autonomous region), whose position is relatively stable, can serve as a linkage in between.

5.4 Summary

The research on the development of national urban space is a complex systemic project. In the context of economic globalization, we must adhere to the basic principles established by the scientific theory of human settlements to achieve sound and appropriate development of national urban space. On this basis, the theory and methodology of national urban spatial planning must be constructed with a broader vision, using the means with higher administrative efficiency, comprehensive instruments to better maintain market functions and compensate for market failures, and more measures to embody social equity.

In the market economy, spatial planning, as a crucial function of the government, is an important means for the state to achieve coordinated economic, social, environmental, and regional development. Its core lies in leading industrial development through the reconfiguration of the spatial structure of towns and cities and realizing the rational use of spatial resources. The technical focus of national spatial planning is on the formulation of urbanization policies and development policies for different regions. In the era of globalization, the urban spatial structure of the country must be dynamic and open, with the flexibility to respond to uncertainty. The effective implementation of national spatial planning must be guided by fiscal and financial instruments. The system of national spatial planning should thus be reformed to establish a system that "integrates the Three Plans into One."

References

1. Niu WY (2006) Three strategic options for China's development model. CPPCC Daily
2. Qiu BX (2005) Changes in urban planning in China's urbanization process. Shanghai: Tongji University Press 92:100
3. Ma K et al (2006) A number of major strategic tasks for China's economic and social development during the "Eleventh Five-Year Plan" period – Interpretation of the Outline of the Eleventh Five-Year Plan for National Economic and Social Development of the People's Republic of China
4. Ye Y, Huang R (2004) Research on the characteristics of China's migrant population and urbanization policies. J Renmin Univ China (2):75–81
5. Zhou YX (2006) The first scientific problem of urban research is the correctness of the concept of foundation stone. Urban Plan Forum (1):1–5
6. Wu C (2003) Theory and methods of territorial planning. Science Publishing House Press
7. Wu LY (2001) Introduction to sciences of human settlements. China Architecture and Building Press, 70, 101:106–112,183
8. Wang K (2003a) Shortcomings of China's planning theory from the perspective of western modern planning theory. City Plan Rev (6)
9. Wang ZL (2003) Investment report of multinational corporations in China. China Economic Press, Beijing, pp 15–23
10. Wang JC (2001) The space of innovation. Peking University Press, Beijing, p 121
11. Friedman J (2005) Life space and economic space: contradictions in regional development. Urban Plan Int 5:8
12. Gottmann J (1957) Megalopolis or the Urbanization of the Northeastern Seaboard. Econ Geogr 33(3):189–200. https://doi.org/10.2307/142307

13. Zhou YX (1986) Suggestions on clarifying the concept of cities and towns and the caliber of urban population statistics in China. Urban Plan (3):10–15
14. Yao S, Zhu Y, Chen Z (2001) Urban agglomerations in China. University of Science and Technology of China Press, Hefei
15. Hall P (2002) Urban and regional planning, 4th edn. Routledge, London
16. Hall S (2002b) The Yangtze River Paradigm. City Plan Rev (12)
17. Wu LY et al (2002) Study on urban and rural spatial development planning in the Beijing-Tianjin-Hebei region. Tsinghua University Press, Beijing pp 3–11
18. Wu LY (2003) Zhang Jian and Nantong as the first modern City of China. City Plan Revi (7):6–11

Chapter 6
Theoretical Considerations on National Spatial Planning

6.1 Revisiting the Role of National Spatial Planning

6.1.1 Reflecting on the Origins of National Spatial Planning

There have been different views on the significance and role of spatial planning at the national level. In 1921, the former Soviet Union initiated national economic regionalization, advocating organized and systematic regional development of the country under the guidance of national plans. Socialist countries in Eastern Europe and Asia (including China) followed suit in the ensuing years, making earnest efforts in productivity distribution and spatial planning. For this, spatial planning at the national level has long been considered the result of a planned economy. After China's shift to a market economy since the reform of the economic system in the 1980s, national-level planning has been neglected for a long time; the abortive planning of land resources in the mid-1980s and the fact that no urban-wide planning has been introduced for 15 years since the promulgation of the *Urban Planning Law* are clear examples. In fact, having a national spatial planning in place or not is only the surface of the problem, at the root of it is whether state intervention is needed for the socio-economic development of a country, and this is a matter of understanding.

Since the 1950s, such capitalist nations as the Netherlands, Japan, and South Korea have been vigorously pursuing national spatial planning or planning of land resources, and the reason for this is worthy of our consideration. From a historical perspective, most of those nations' planning were done during their economic take-off, and their primary goals were to spur rapid economic development and grow national strength in a relatively short period of time. It is a straightforward path to development to gather, through state intervention, the limited financial resources for developing certain regions first, creating the growth pole so to drive the development of other regions and the overall national economic development. Adjustments are then made to the planning for the following reasons: break excessive polarization of certain cities so as to create a relatively balanced regional spatial pattern;

K. Wang, *National Spatial Planning in China*, Urban Sustainability,
https://doi.org/10.1007/978-981-97-7729-7_6

given the importance of resources and environment—a lesson drawn from the oil crisis, propose spatial development plans taking account of the limited resource; and realizing the importance of national and regional competitiveness in economic globalization, actively pursue macro spatial planning, and so on. This indicates that the formulation of national spatial planning is driven not only by the needs of spatial development itself but also by considerations of national development strategy.

The introduction and development of national spatial planning does not entirely depend on the economic system; national planning exists too under a market economic system, and the key here lies in how to understand the relationship between the market and the government. Some scholars believe that "government" and "market" are like two "wheels" working together in promoting social development. The question lies not in whether there is "strong government intervention", but rather at what areas does the "intervention" target and with what goals [1].

6.1.2 Enlightenment from the Development Dialectics of Underdeveloped Countries

The study of the modernization process of underdeveloped countries has been a focus of international research. Two of the theories that have considerable influence are G. Myrdal's Theory of Circular and Cumulative Causation and A.O. Hirschmann's Theory of Non-equilibrium Development. According to the former, in the economy of underdeveloped countries exists a dual structure in which developed and underdeveloped regions coexist; the former have "echo effect" and "diffusion effect" on the latter. And due to the influence of market forces, the echo effect has far greater influence than the diffusion effect, rendering the developed regions even more developed and the underdeveloped regions left further behind. In here lies the root cause of imbalance of regional economic development. And for this, Myrdal believes that it is necessary for the government to intervene for the coordinated development of regional economy. When certain regions have attained a good momentum of development, the government should adopt an uneven development strategy, giving priority to the regions with strong growth potential, with the aim of achieving higher investment efficiency and rapid growth rate, and boost the development of other regions through diffusion effect. The Theory of Non-equilibrium Development, on the other hand, believes that development is a chain of unbalanced evolution with strong forces of economic growth concentrating on the space around the initial starting point. And the emergence of growth poles means that regional inequality in growth is both an inevitable result and a prerequisite. Therefore, theoretically speaking, it is a general rule that with underdeveloped countries the path to modernization should be from unbalanced to balanced.

It was in 1949 when the New China was founded that the modern industrialization of the country began in earnest. However, the development from 1949 to 1978 was hard and tortuous. The glory of the 1st Five Year Plan gave way to various political movements such as the "Great Leap Forward" and the decade of the Cultural Revolution. By 1978, China's per capita income was at the bottom of all the countries, rendering it a low-income country. According to the statistics of the World Bank, and based on the PPP international dollar value (current price), China's per capita income in 1978 was 340 USD, nearly half of the average 660 USD of low-income countries, 26% of the average 1290 USD of lower-middle-income countries, 10% of the average 3140 USD of upper-middle-income countries, and 4.1% of the average 8180 USD of high-income countries. It was against this backdrop that the Reform and Opening up proposed the uneven development strategy for the first time. To quote Deng Xiaoping's words in 1978: "I think we should allow some regions and enterprises and some workers and peasants to earn more and enjoy more benefits sooner than others, in accordance with their hard work and greater contributions to society[1]." In 1980, he again pointed out that "we should acknowledge the existence of imbalances, and there is no hope of engaging in equalitarianism. Only by allowing some regions to become rich first can the country gain extra strength to help the regions left behind[2]." Several spatial strategies were proposed, including the establishment of special economic zones in the provinces of Guangdong, Fujian, and Hainan, opening 14 coastal cities, and opening the areas along the Yangtze River and the border. Remarkable progress has been achieved through the implementation of these strategies, as evidenced by China's per capita GDP reaching 1000 USD by the end of 2003 and the sustained, rapid economic growth over the years.

Nevertheless, the strategy of uneven development also brought about its unique problems: the gap between urban and rural areas has widened, with the comparative value of per capita net income of rural residents and per capita disposable income of urban residents reaching 1:3.2 in 2003, and the income of rural residents have stayed the same or even declined for eight years. The huge difference between urban and rural areas has led to a widening gap between the eastern, central, and western regions, resulting in a serious imbalance in regional development. A large amount of rural labor force has shifted from the central and western regions to the eastern region, and the urbanization rates of the three major regions vary greatly, with 44.60% in the east, 33.50% in the central, and 27.66% in the west, a gap of nearly 17 percentage points between the east and the west. So is the same with the degree of marketization in the three regions, with 80% in the east, 50% in the central, and 30% in the west. The degrees of the central and western regions are lower than the national average. In fact, the gap in terms of urbanization and marketization is only the phenomenon, what lies beneath it is the imbalances in wealth distribution and interest relations. And these imbalances have affected the coordinated development of national economic and social structures. Comparing Iran's "White Revolution" in the era of Pahlavi in the 1970s that advocated modernization with the failure of Argentina's "Second Revolution" during the reign of Menem that upheld defining national development strategies in light of globalization, the study believed that the political life of a country

[1] From the speech "Emancipate the Mind, Seek Truth from Facts and Unite as One in Looking to the Future" delivered by Deng Xiaoping on Dec 13, 1978.

[2] Remarks made by Deng during his visit to Wuhan in mid to late July 1980.

must see to it that coordinated and balanced development adapt to the economic development in modernization, which should be the inherent meaning of political equilibrium and also a prerequisite for economic modernization [1].

Since the Reform and Opening Up, China's development policies shifted from the gradient development strategy centered on non-balance in the 1980s to promoting all-round, balanced and sustainable development of the economy and society characterized by relatively balanced development at the third plenary session of the 16th Central Committee. The shift embodies the new leadership philosophy of using politically balanced means to "redress" the uneven consequences of economic modernization and "steer" its course, i.e., by providing more complete, abundant, and equitable balanced public goods to maintain the vitality of economy and society through moderate non-balance in the economic field while ensuring the coordinated development of the economy and society via politically balanced means [1]. In short, the unbalanced strategy is a necessary step in the modernization of underdeveloped countries, its aim being paving a solid economic foundation. The balanced strategy is the long-term goal and political need of national development. All this is for the sake of development. Based on the development and current situation since China's Reform and Opening Up resorting to spatial planning for moderate intervention in the imbalance of development is born out of the need of both development as well as national political equilibrium. The newly released 11th Five Year Plan emphasizes the importance of coordinated regional development and the management of spatial resources and the spatial distribution of cities and towns, reflecting the central government's increased awareness of the significance of spatial intervention in the new situation and is taking action to implement it.

6.2 Rethink on the Significance of National Land Space

6.2.1 General Meaning of Space in the Era of Globalization

Urban and rural planning is, in the final analysis, about the management of space. Therefore, rethinking the meaning of space at the macro level can help us orient planning. In his Cities in Evolution published in 1915, Patrick Geddes first used the concept of region to analyze and understand cities, predicting that the expansion of cities in regions is inevitable and will form a new spatial system. He used the concepts of "city region" and "conurbation" to describe this feature. C.A. Doxiadis expanded the scope of space even further. From the perspective of human settlement, he proposed 10 levels of settlement, including the analysis of city-to-metropolis, urban continuity, urban continent, and ecumenopolis, providing a new interpretation of the structural system and significance of urban space.

With the cross-border flow of capital in the era of globalization, the global economy is seen as a "space of flows" [2], and regional space is generally deemed a geographical space without borders. After the Second World War, the state-based accumulation of capital and political regulation has been gradually expanding to a larger scope worldwide. Taking Europe as an example, in the past 50 years, with the advancement of the European integration process, the political and economic role at the regional level has been steadily strengthened. The establishment of both the single currency system and the structural funds indicate that transnational regional coordination mechanisms are playing an important role. And the successful adoption of the ESDP in 1999 is significant in that it provides a transnational economic and social development framework. Amin believes that the globalized economy is synthesized in different places, at different levels, and in different ways. Region is an important category, characterized by the uneven distribution of different regions in the international division of labor, as well as the assimilation and alienation of different spaces [3]. The globe is "composed of places" [3], and the places are "projections of globalization". Against the backdrop of globalization, the meaning of space is twofold: "home" and "world". Space is fluid and changing, and the original hierarchical structure is discrete and diverse.

6.2.2 The Uniqueness of National Land Space

There is no consensus on the significance of national land space in the era of globalization. Most views are that the role of national boundaries has weakened considerably, even disappeared. Most of these views and theories are based on the understanding of economic globalization, that is the economic significance of space is fully demonstrated. However, the "9/11 Incident" in New York, United States, in 2001 raised some doubts about the concept of space without borders [4], namely, the political significance of land was reconsidered and recognized. This is proof that in the past few years, our understanding of geographical space has indeed been limited to the scope of "economic geography". In the analysis of national land planning, more weight is given to such technical aspects as productivity distribution than considerations of national development strategy.

In fact, for developing countries, in addition to bringing benefits such as capital and job opportunities, the globalization theory sees to it that they will always be in the middle section of the industry chain, that is, the part with the lowest added value, and always at the outermost ring of the core of technology. In other words, the status of underdeveloped countries in economic globalization has not improved, on the contrary, they are often marginalized. The most extreme example is the "Second Revolution" in Argentina. After taking office in 1989, the Menem Cabinet, in the face of globalization, proposed the globalization theory characterized by the behavior mode of "political subordination", namely, it is politically compliant with the hegemonic position of the United States and economically sought to fully integrate with the world economy. It had then abandoned the path of industrialization it had been pursuing and placed the focus of the national economic development strategy on the "specialized production" of agricultural and animal husbandry food. Joining the international division of labor system as a single exporter of primary agricultural and animal husbandry products has resulted in its own industrial system vanquished due to trade liberalization, and gradually the country was trekking on a path of development leading to poverty and dependency, and ultimately political instability[3]. Some scholars stated that the main reasons for Argentina's failure in globalization are: first, it made the mistake of abandoning the principle of independent development and fell back to depending on others; and second, it failed to handle the relationship between the market and the government, namely, it is inevitable that a market economy leads to inequality in the distribution of social resources and political inequality among citizens. In the context of globalization, due to intensified competition, the existing national welfare system was disrupted, leading to the escalation of social conflicts [5]. Therefore, in the era of globalization, national land space, possessing political, economic, and cultural values, cannot be simply understood from the economic perspective, and the development of a country has a political value that we can ill afford to neglect.

Theoretically, the understanding of the political significance of space began with David Harvey. He transformed the neoclassical economic geography into Marxist political and economic geography, a shift that turned the study of economic geography from being limited to the statistical laws of spatial distribution and the understanding of industrial location to the analysis of the historical process of regional development and the inherent "laws of movement" and the "crisis tendency" of capitalism. As a result, western geographers were prompted to conduct extensive analysis of regional and urban issues, including dynamic analysis of urban space and regional development crises, analysis of the process of labor transfer, analysis of regional industrial composition, and analysis of international unbalanced development. For more than 20 years from the 1980s to now, the mainstream theory in the geographical field is still based on the spatial analysis of economic laws; even Porter's theory of national competitiveness too is based on such economic concepts as "enterprises cluster". Therefore, although technically we have entered the postindustrial or information era, moving towards an economy with services, knowledge, and information as the core, national development, however, has not been simply integrated into the so-called virtuous circle of the globalized economic trajectory, rather, it is constantly undergoing alienation with a strong political and cultural color. Otherwise, the tragedy of Argentina would not have happened, and the frequent trade frictions between China and Europe and North America would have been avoided too. Therefore, the significance of space in the era of globalization possesses not only the quality of "space of flows" resulted from capital flows, but also that of "fixed space" due to the country being the unit of political and culture value. And in a world where ideologies still prevail, the significance of the latter is more relevant.

[3] This part is paraphrased according to *Political Logic of Balanced Development* by Hu Weijun.

6.2.3 The Role of Space Governance

Hall believes that the essence of "Garden City" is local autonomy and self-management founded on urban-rural coordination [6, 7]. He believes that the global economic recession beginning in the 1970s has prompted Western planners to rediscover the role and power of the market. New practices of "Development Corporation" and "Enterprises Zone", which do not comply with the original planning system, were widely adopted in the 1970s (first in the United Kingdom, then in other countries, including China) precisely because the primary goal is to promote the development of local economies regardless of how planning decides the form or development of cities. With heightened international competition, local governments need foreign capital. Some Western scholars believe that introducing foreign capital and increasing job opportunities are also the goals and objectives of planning [8]. Hall also believes that the reason why planning still works today is that there are still enterprises that require collective action, for example environmental protection [9].

Friedman [10], on the other hand, believes that in geography, dwelling space and economic space together constitute a unity of opposites. In the past two centuries, economic space has been disrupting, invading, and dissolving the dwelling space of individuals and society. Economic space is abstract and intermittent, a space having nothing to do with history. When economic hegemony prevails over life, development is simply understood as the expansion and growth of production, regardless of social costs. He proposes that life must come first, with livelihood issues taking the second place. The country must take control of the basic conditions of livelihood in certain areas and fully exercise its power in areas that have a bearing on the fundamental interests of the people [11]. From the perspective of human well-being, he believes that the so-called well-being based on per capita wealth is neither comprehensive nor objective. Many issues in regional development such as the dismemberment of a large amount of living space and deterioration of the ecological environment call for political solution of communication and consultation.

In recent years, the study of urban and regional governance has become a popular theme in planning theory. Theoretically, governance is about exploring the balance of rights among various social forces, while regional governance refers to the power interaction between different levels of government and development entities, as well as between governments at the same level [12]. Regarding the governance of regions, there are different perspectives on centralization and decentralization. The decentralization theory emphasizes local autonomy, believing it can promote more participation in political life and social management of the people. The theory of centralization focuses on the important position and role of the central government in the overall social coordination and control, believing it can bring about social production and eliminate anarchist competition, which is conducive to the sound allocation of social resources. The "new public management" or "enterprise managerialism" emerging in recent years is a compromise between the two. Although decentralization has always been the mainstream tradition of regional governance in large cities in the West, in the era of economic globalization, regionalism became popular once again.

Seeking the overall interests of regions and conducting coordinated management among regions has become a new concept of governance.

To sum up, the management of regional space is in essence to improve the development capacity and quality of life of the region. The liberalization of market mechanisms does not mean that social life can be neglected. From the perspective of the constituent elements of regional space, a region cannot have economy as its single attribute, and the governance of space is needed for the protection of comprehensive interests.

6.3 Rethink on the Role of Planning

6.3.1 Modern Planning Theories—From Control to Guidance and Communication

Since the publication of Ebenezer Howard's *Tomorrow: A Peaceful Path to Real Reform* in 1898, theoretical and practical exploration has been continuously carried out on the control and guidance of urban development, and the scope of research has also been expanding. The theory of "Garden City", Abercrombie's Greater London Plan, the Netherlands' national spatial planning, and Japan's national land planning have left a deep impression on people, helping establish the control concept for urban and regional development. The regional planning and national land planning that we are discussing today can trace their theoretical root to urban planning which has a close bearing on regional planning and national land planning. Charter of Athens, formulated in 1933, states that "every city should prepare an urban planning program that is consistent with national planning and regional planning."

Since the 1960s, the major planning theories of the West underwent stages of changes under their continuously changing socio-economic circumstances. On one hand, they launched fierce attacks on the traditional theories based on technology rationality. On the other, with the progress of globalization, their understanding of the scope and significance of space management has also been constantly updated. In the past 40 years, the planning theory has been through the following major stages:

Systems Theory: introduced by Brain McLoughlin and George Chadwick in the 1960s. The systems theory of biological science has been fully employed in the analysis of cities. Biology believes that systems exist in various fields of the natural and human environment and can be controlled through planning[4]. The core of the method of system planning is that cities and regions are interactive complexes. The system is constantly changing and flowing. As a form of system analysis and control, planning itself must be dynamic and changing. Planners must find ways to categorize and predict these changes to control them. McLoughlin pointed out that planning seeks

[4] According to the definition of the Oxford English Dictionary, a "system" is a "complex whole", a "set of things working together as parts of a mechanism", and a "group of related parts that work together as a whole for a particular purpose".

to regulate and control individual and collective behaviors, minimize bad benefits, and promote the realization of a good physical environment considering the general and specific objectives of the planning.

Rational Process Theories of Planning: proposed by Andreas Faludi in his *Planning Theory* published in 1973. The theory emphasizes using "scientific" and "objective" methods to understand and plan cities, and planning as a method is all about producing the best results [13]. Like engineers, planners should seek out the best methodology and arrive at a comprehensive and rational choice by considering various special requirements and solutions. Rational planning first defines the objects to be planned and the objectives of planning, analyzes them, break them into a set of problems and objectives, selects the best solution, and constantly checks the results. Emphasis is placed on acting according to procedures, continuous decision-making, and comprehensive review. That is in terms of methodology, it emphasizes the defining of goals, problem sorting, and comprehensive decision-making. According to Faludi, the process of rationality begins with identifying and defining the "problems" raised, and in the process, it is necessary to distinguish which is the rational one.

Western Marxism and Critical Theory: proposed by David Harvey in the 1970s. The theory holds that cities and planning (including theories of planning) are reflections of capitalism, helping to construct capitalism. In capitalist society, there is no so-called "public interest", only the interests of capital. The interests of capital form a national mechanism through such means as planning to achieve public control. Western Marxist theory is divided into the following major categories: the state as a tool for the ruling class, the state as the provider of necessary intervention, and the state as a cohesive factor. Among them, "the state as the provider of necessary intervention" has the greatest impact on planning. In fact, Keynesianism is founded on the theory that capitalism can be "controlled", that is, development is controlled through the relationship between supply and demand, and urban planning is another mechanism of national control to rid of the economic crisis. In Western Marxist theory, the state plays the key role. Planning serves as an extension of national interests, reflecting the needs of capital through orders. Thus, this also creates the necessary conditions for providing certain forms of state intervention (land, capital). State intervention is particularly needed for such public infrastructures as roads and bridges because while requiring large amount of investment, they generate little revenue. In a market system, most planning is in fact oriented towards the requirements of market, and there should be more strong opposition to market forces [14].

New Right Planning Theory: the new right theory was born in the 1980s, with Hayek, F and Friedmann, J as representatives. The theory is a combination of competition through market orientation (liberalism) and strong government intervention (conservatism). The core of the new right of liberalism is the dominance of the market. In his book, Hayek states that central planning is (although not entirely) dangerous and ineffective, it interferes with the market, restricts individual freedom, and erodes the foundation upon which to establish the principle of prudent law in the state machinery. The interaction in the markets leads to natural order and stratification of society. Planners cannot hope to replicate society by themselves, as they know only a small part of society. Government and state intervention should be limited in scope, for example to the extent of law, infrastructure, and national defense. According to the new right perspective, planning should only be carried out at the local level, and market planning is the best foundation for all decision-making. The conservative New Right puts its emphasis on the role of strong government. Due to the emergence of crimes, lack of order, and vandalism, etc. in Western cities after the war, the conservative New Right believes that order can only be maintained through a strong government.

Pragmatism: a trend of planning in the late 1980s. Its representative figures are Dewey, J and Rorty, R. It emphasizes the direct solution of special problems under certain conditions and situations. This trend of thought is mainly prevalent in the United States, and it has made great progress in the past 20 years. A relatively complete summary of pragmatic planning is given by Charles Hoch. He believes that in practice, experience is a better arbiter than theory; answers born out of practice is facing the real questions; and the practical approach should be achieved through social consensus and democratic means. Therefore, planning should be based on communication and planners play the role of gatekeepers, whose task is to choose from multiple possibilities.

Planners as Advocate Theory: originated in the 1970s and its representative are Dannis, N, and Davidoff, P. It upholds non-neutrality of planners, believing that planners should participate in the political process as representatives of government interests, group interests, organizational interests, or individual interests of the communities that may be affected by policies. It believes that planners can enable the public to play a real role in the democratic process. Some scholars believe that the essence of advocate planning is pluralism. Decentralization makes it possible to change the original meaning of policies during its formulation through the involvement of different departments. The meaning of political decisions is constantly changing during implementation and by different departments executing them. In a pluralistic system, all participants would return to the system during the bargaining of uncertainties.

Postmodern Planning Theory: originated in the mid-1980s and represented by Sandercock, L and others. The attention it garnered today has much to do with the rising of the information society. Webster's definition of the information society include the following five aspects: in terms of technology—the computer technology permeates the society in its every pore, effecting social changes; in terms of economy—information economy relates not only to the information technology manufacturing industry, but also the contribution of the application of information technology to GDP growth; in terms of changes in occupation—in the information society, job opportunities in manufacturing are decreased in large numbers while that of service industries are increasing; in terms of spatial change—thanks to the development of information networks, regions that are previously far away from each other are connected, resulting in the "time-space compression"; in terms of culture—a large amount of high-quality information are accessible through various media such as television, PC, and personal communication, indicating the advent of the information society. As far as postmodern planning is concerned, Sandcock has defined five new principles: social justice is as important as market effectiveness;

the definition of a problem by different political groups should be reached through discussions among those groups; an inclusive moral value should be established; the ideals of the community; public interests should move towards a more diverse and open "civic culture".

Collaborative Planning Theory: Representatives include Habermas, J and Foucault et al. Its main idea is to see planning as a process of communication and collaboration. The theory represents a partial return to the rationality of modernism. Dryzek believes that breaking the shackles of interest groups and fostering a "true public atmosphere" allows people to have objective discussion and reach broad consensus on certain issues of common interest, the typical examples are Greenpeace, Eco organizations, anti-nuclear organizations, civil rights organizations, etc. He believes that communication rationality has the capacity to bring about a complete set of procedures for political organizations. As a communicative approach to planning, Forester proposed the following set of working methods for planners: cultivating networks for social exchange and communication; paying special attention to the interests of those who do not have an organization as their supporter; education of citizens and community organizations; improving the skills of planners to work with other groups; encouraging independent reflection on community-based projects, etc. In short, collaborative planning is a negation not only of the absolute truth held by instrumental rationalism, but also of postmodernism which denies the objective rationality of objects.

6.3.2 Rethink on the Role of Planning

By the induction and analysis above, we can see that although there are many different modern planning theories, the main ideas are rationality, comprehensiveness, state intervention, market discipline, and broad communication. Although requirements for planning differ in spatial planning at different levels, the core of planning is ultimately the management and guidance of space. From the most original concept of planning, Howard's "Garden City" theory is a rational arrangement of urban space, only this arrangement emphasizes the autonomy of people and human freedom in the space.

We can also appreciate the significance and role of planning by analyzing the definitions of the various plannings[5]:

(1) "As a common activity, planning refers to the preparation of a coherent sequence of actions to achieve the set goals" [15].
(2) "Planning is the formulation of a set of decisions to determine future actions, that is, to help achieve goals in the best way, and to learn from the results new possible decisions and the new process of pursuing goals".

[5] This part is based on *The Philosophy of Urban Planning* by Sun Shiwen. Beijing: China Construction Industry Press, 1997. 18

(3) "Planning is essentially an organized, conscious, and continuous attempt to choose the best method to achieve a specific goal" [16].
(4) "Planning is the rational application of human knowledge to the process of achieving decisions that will serve as the basis for human action" [16].
(5) "Planning is an effort collectively determined through democratic mechanisms and the goal is to make concentrated, comprehensive, and long-term predictions about future trends... to propose and carry out coordinated policy systems that are designed to connect foreseeable trends and achieve desired outcomes, and to articulate concrete goals in advance".

We can see from the various definitions that "rational arrangement" of human actions is fundamental to planning, otherwise everything can be left to the market to determine. Under different economic systems, the role of planning varies greatly: under a planned economic system, spatial development becomes the continuation and concretion of planning, as a result, its role of democratic and pluralistic coordination is not fully played. Under the market economy system, the market plays a fundamental role in the allocation of resources, and the government serves an institution that intervenes only to safeguard the interests of the entire society, making up for what the market fails to provide. Government interventions are materialized by way of policies and such means, and planning itself is a tool embodying policy [9]. Therefore, regardless of the various arguments in its definition or methodology, the role of planning, in the final analysis, lies in the rational application of human knowledge to the decision-making process; those decisions will serve as the basis for human action. With the development of society and the acceleration of the process of globalization, acknowledging the objective laws of the market in the era of globalization while giving play to the role of the government in macro control to make up for the flaws of the market is the prerequisite for establishing the urban planning theory of China.

In the context of a market economy system, the significance and role of national spatial planning in China pose a thought-provoking theoretical question. This proposition involves two crucial premises: first, whether state intervention is necessary under a market economy system, and second, whether national spatial planning can contribute to national development. The former question has been affirmatively addressed in previous theoretical analyses, while the latter has received preliminary answers through international comparative analysis. However, the effectiveness of national planning in China's modernization process depends on its alignment with the core goals of national development.

Considering the current development status and the direction of major policies in China, breaking through resource constraints and achieving balanced development are strategic imperatives in the present and foreseeable future. In terms of resource conditions, China's per capita resource ownership is significantly lower than the global average. Per capita land, freshwater, forest land, grassland, mineral, and marine resources account for only 33%, 27%, 14%, 50%, 58%, and 25% of the world per capita average, respectively. Coal, oil, natural gas, and hydropower resources per capita represent 70%, 10%, 4%, and 63% of the world per capita average, respectively.

Therefore, it becomes essential for national spatial planning to fully embody the goals of land conservation, water conservation, and energy conservation at the national level. Simultaneously, addressing the imbalances that have persisted over the past two decades of development (as discussed in previous sections) becomes a key focus of national spatial planning.

6.4 Understanding the Role of National Spatial Planning in China

6.4.1 Requirements of National Development Objectives

The relevance and role of national spatial planning within China's framework of market economy represents a theoretical inquiry that warrants profound contemplation. This proposition entails two pivotal tenets: firstly, the necessity of governmental intervention within the context of market economy; and secondly, the potential of national spatial planning in fostering national development. The former query finds affirmation in the above theoretical analyses, while the latter is addressed on a preliminary basis within the realm of international comparative studies. However, the efficacy of national planning in China's modernization process hinges upon its alignment with the core objectives of national development.

Given China's current development context and overarching policy orientation, it is a strategic imperative to break away from the constraints of resources and strive for balanced development for the near and distant future. Considering its resource endowments, China's per capita allocation of resources markedly trails behind the global average. For instance, the per capita availability of land, freshwater, forested areas, pastures, minerals, and marine resources registers at a mere 33%, 27%, 14%, 50%, 58%, and 25% of the global average respectively. Likewise, the per capita shares of coal, oil, natural gas, and hydro-power resources amount to 70%, 10%, 4%, and 63% of the global average. Hence, the imperative lies in thoroughly integrating the objectives of land, water, and energy conservation through national-level spatial planning. Furthermore, addressing the historical imbalances in development over the previous two decades (extensively elucidated earlier) stands as a key facet of national spatial planning.

The then Premier of the State Council Wen Jiabao once underscored that "Establishing a scientific outlook on development is exceedingly critical, considering the international experience and array of conundrums confronting our nation as it enters a fresh stage of economic and social development. Throughout the years, China has accumulated a plethora of contradictions and challenges alongside its rapid economic progression, mainly stemming from the widening urban-rural gap, regional disparities, income divides among residents, mounting pressures on employment and social security, laggard development in education, healthcare, culture and beyond, exacerbation of tensions between burgeoning population growth and economic development and ecological environment and natural resources, backward growth models, and the general lack of quality and competitiveness in its economy. Addressing these contradictions requires serious attention rather than evasion. It entails a gradual approach to its resolution instead of business as usual. This juncture presents a watershed moment in China's modernization process and symbolizes a pivotal stage where economic and societal structures undergo profound restructuring. The developmental trajectories of numerous countries indicated that at this critical phase, two divergent outcomes are plausible: a favorable trajectory with ongoing economic and societal progress and successful industrialization and modernization; or an adverse scenario beset by disparities between the affluent and the impoverished, increasing unemployment, widening urban-rural and inter-regional gaps, exacerbation of social contradictions, and ecological degradation—a trajectory often marked by protracted economic and societal stagnation, or even turmoil and retrogression. "The excerpt from Wen's speech here intends to elucidate that national spatial planning goes beyond a mere spatial tactic for urban development; instead, it embodies a tangible roadmap to implement national strategy. Its control, guidance, coordination, optimization and safeguard mechanisms stand as pivotal components of its macro-regulatory function.

6.4.2 The Theoretical Core of "National Spatial Planning Theory"

Based on the comprehensive analysis above, the national land is first and foremost multifaceted space, with economic, social, and ecological dimensions as its main components. Consequently, it is impossible to have one single target or aspiration for spatial development. National spatial planning, as a multifaceted subject of great complexity, involves resources, environment, population, industry, and systems, among other factors. It is key to strike a balance between development facilitation and resource constraints. This intricate interplay requires research through the lens of implementation of national strategies and facilitation of social-economic development. It also necessitates efforts to optimize the overall spatial development of national land based on sustainable development objectives of urban areas and conservation and reasonable utilization requirements of natural resources.

Emerging as a latecomer to urbanization, China has condensed the 200-year urbanization journey witnessed in Western countries into merely 40 years since its reform and opening up. It managed to materialize the largest urbanization ever seen in the world. The urbanization rate has surged from 17.9% in 1978 to 63.9% in 2020, along with expansion of urban construction land from 11,600 square kilometers in 1990 to 58,600 square kilometers in 2020. This highly condensed spatial and temporal evolution, while propelling rapid economic and societal advancements, has triggered a series of challenges such as ecological and environmental degradation, pronounced hazards, and the fading allure of cultural characteristics. At the heart of these dilemmas is the single-handed prioritization of economic prosperity in urban distribution at the expense of natural and cultural factors. It fails to align with the diverse natural inventories and cultural characteristics across different regions of China. Hence, given China's complex national realities defined by geographical diversity and great variances in natural inventories and cultural characteristics, it has become a major scientific quandary in national and regional spatial planning to strike a balance between development and security priorities and strive for urban layout adaptation to natural and cultural conditions. In response, the national spatial planning theory is a theoretical approach to planning based on the urban spatial development research that precedes it, and the science of human settlement. It centers around the core idea of "precise adaptation" of cities and towns to natural and cultural conditions. The "precise adaptation" refers to coordinating urban areas (the core vehicle for economy and society) with natural and cultural conditions at national and regional levels. It aims at the accurate, rational and adaptive alignment of the urban layout with facilities and differentiated and dynamic natural environments, cultural characteristics, and socio-economic development in different regions. It mainly includes the following theoretical connotations.

First of all, building on the aforementioned technical framework for urban spatial development, a layer-by-layer analysis on spatial resources has been proposed, with ecological security as the prerequisite. It takes stock of three layers of spatial resources and divides them into ecological security layer (featuring natural environment), infrastructure layer and human settlement layer(featuring urban construction). The ecological security layer refers to the spatial resources based on natural and ecological elements, which is the foundation of human settlements; the infrastructure layer is the infrastructure network with transportation as its core, which is the skeleton of the space; and the human settlement layer includes the varying types of human settlements at different levels, which is the core of human life. The "national spatial planning theory" elaborates on the spatial elements of the three layers from the lens of pleasant living environment. That is, it takes the ecological security layer as the core of spatial resource analysis and the basis of urban layout, evaluates the carrying capacity of regional spatial resources by quantitative analysis and takes into account the trends of population mobility and industrial development analysis. It studies and identifies the development trends of urban areas, so as to create an urban spatial distribution and infrastructure network that is compatible with the natural environment, realizing the adaptation of urban layout to natural environment.

Secondly, in view of the diverse and disparate regional spaces in China and the uncertainty brought by the intensification of global climate change, an effective means to ensure the spatial security and sustainable development of the national territory is to figure out the resource inventories in different regions, create an adaptive urban spatial distribution and continue to monitor, evaluate, and optimize the dynamic changes in space. Therefore, the "national spatial planning theory" further proposes a full-cycle regional spatial research method and framework, featuring "precise analysis, adaptive distribution and dynamic evaluation". It means creating urban areas that are compatible with disparate and dynamic natural environments in different regions, so as to provide vital support for the sustainable development of urban areas.

6.4.3 The Expansion of "National Spatial Planning Theory"

In the context of China's further urbanization, exacerbating global climate change and evolving problems in national, regional and urban development in recent years, the connotations of "national spatial planning theory" have been constantly expanded and improved. During China's urbanization process over the past 40 years, the first 30 years saw a rapid development of cities and towns. In the face of tightening resource and environmental constraints, severe environmental damage, outstanding land supply and demand conflicts, and resource depletion, the core issue at this stage was to balance the need for rapid development and the goal of sustainable development. Therefore, during this period, the "national spatial planning theory" centered on regional spatial resource analysis based on natural and ecological elements to create adaptative urban spatial distribution focusing on ecological concerns.

In the past decade, as urbanization slowed down and global climate change intensified, the problems that accumulated in the first 30 years under the high-density and high-intensity development models have burst out, exposing the inadequate abilities of regions and cities in addressing disasters and posing severe challenges for the safety of human settlements. Therefore, the "national spatial planning theory" has shifted the focus to safety issues and strived to adapt urban spatial distribution to cope with natural disasters in different regions by analyzing regional safety risks and identifying their types and risk levels.

In addition, on top of the natural and environmental elements such as ecology and safety, cultural characteristics such as landscape characteristics and native culture are an indispensable part of a beautiful living environment and the high-quality development of national space. Therefore, combined with real-life projects, the "national spatial planning theory" proposes to analyze the cultural resources in regional space and create an urban spatial distribution and cultural space system adapted to the cultural characteristics in different regions.

At this point, with the "precise adaptation" of urban areas to natural and cultural conditions as the core, the "national spatial planning theory" has developed a full-cycle theoretical framework of "precise analysis, adaptive distribution and dynamic evaluation." It consists of two layers, namely ecologically sound and safe natural environment and cultural characteristics, serving as a theoretical foundation for building the technical system of national and regional spatial planning (See Fig. 6.1).

6.5 The Spatial Planning Technical System Under the "National Spatial Planning Theory"

Drawing upon the theoretical framework of "National Spatial Planning Theory," a full-cycle regional spatial optimization technical system comprising three key techniques: precise analysis, adaptive distribution, and dynamic evaluation has been developed (see Fig. 6.2). It combines various tools and methods, promotes the efficient integration of big data, information technology and planning technology and creates a more comprehensive deductive model for spatial analysis and optimization. It also provides new tools for regional spatial optimization.

"Precise analysis" indicates to carry out precise analysis of various factors, including ecological, security and cultural factors, in regional space in a differentiated and targeted way. It includes static analysis of the current status of regional space, and dynamic analysis and prediction of historical changes and possible future scenarios. "Adaptive distribution" means to create an urban spatial distribution and infrastructure network that is adapted to nature, culture, and socio-economic development by analyzing the population and industrial development. Instead of a development blueprint under the ultimate goal, spatial planning is about multi-solution simulation based on various target scenarios. "Dynamic evaluation" means to monitor and evaluate and optimize the development of regional space in a timely manner, so that the urban space continues to adapt to natural and cultural conditions.

6.5.1 Precise Analysis Technology

"Precise analysis" refers to the formulation of a multifaceted evaluation index model, with natural ecology, safety risks and cultural characteristics dimensions. This model enables differentiated and targeted analysis of the resource elements of national and regional spaces, so as to accurately identify suitable spaces for construction, discern safety risk categories and recognize spaces with cultural characteristics in national land space.

(1) Precise identification of suitable construction space

Building on the natural ecological dimension, the suitable construction conditions for towns and cities are thoroughly evaluated through four components: evaluation of basic land use conditions, human settlement environment factors check, safety risk factors check, and resource carrying capacity check. The evaluation method integrates static suitability analysis with dynamic spatial expansion simulation. The combination of static analysis and dynamic simulation is used to accurately identify "pleasant" and "non-pleasant" living space in the national land space (Fig. 6.3).

Evaluation of basic land use conditions To calculate the topographic slope with full-area DEM, and grade the suitability of slopes for urban construction based on gradients of ≤3°, 3–8°, 8–15°, 15–25°, and >25°[6]. The basic land conditions for urban construction are classified as suitable, fairly suitable, generally suitable, fairly unsuitable, and unsuitable. Currently, the resolution of full-area DEM data

[6] With reference to the Technical Procedures for Investigation of Current Status of Land Use, the Specification for Vertical Planning of Urban Land Use (CJJ82-99) and other relevant specifications. According to the basic conditions of urban construction in China, a comprehensive approach is adopted to determine the gradation of slopes suitable for urban construction.

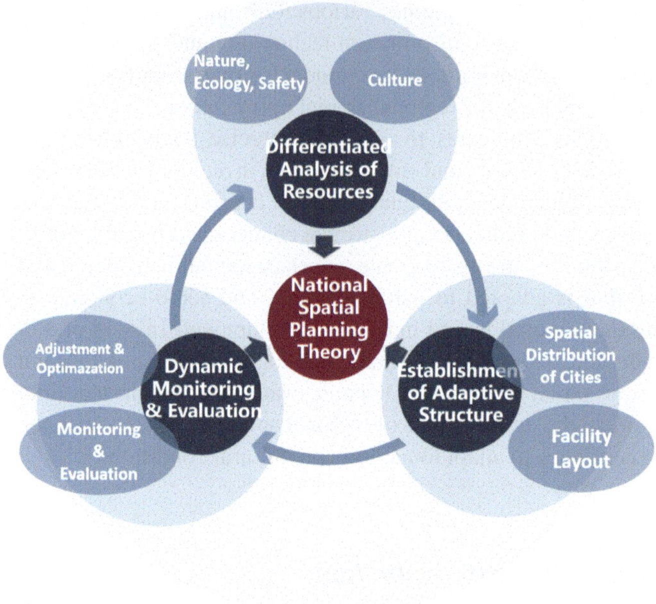

Fig. 6.1 Theoretical diagram of national spatial planning theory. *Source* Prepared by the author

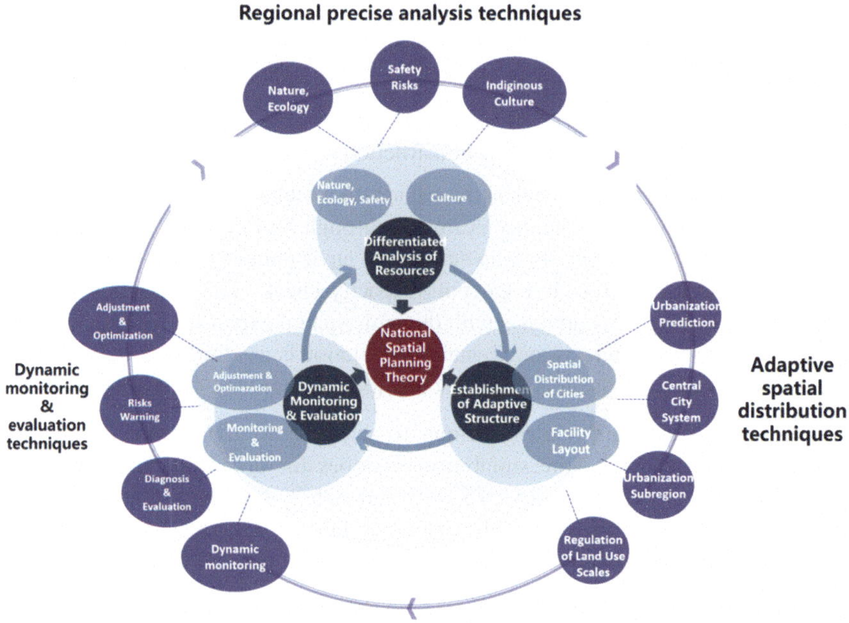

Fig. 6.2 Technical diagram of regional spatial research. *Source* Prepared by the author

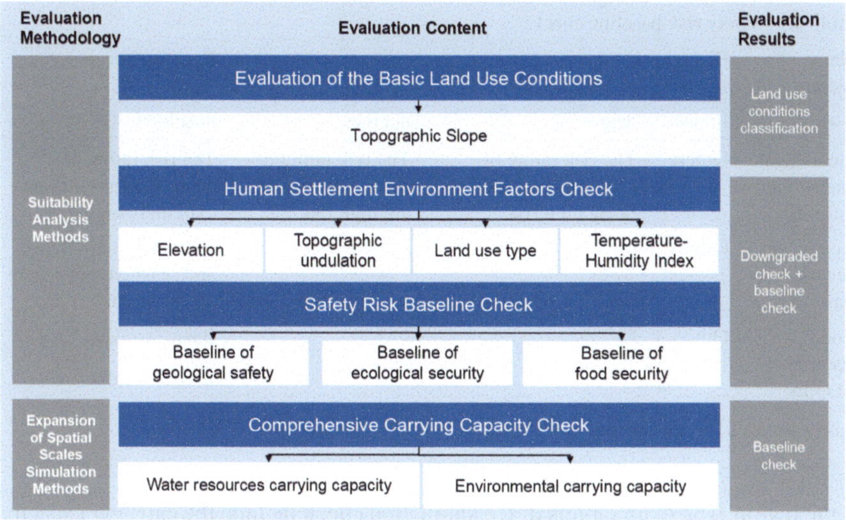

Fig. 6.3 Precise identification process of suitable space for construction. *Source* Prepared by the author

extends to the meter scale, enabling the selection of varying resolution data tailored to distinct evaluation accuracy standards across different scales of analysis for modeling purpose.

Human settlement environment factors check Taking into account the impact of natural and environmental conditions on human life and production activities, the human settlement environment factors such as topographic undulation, land use type, elevation, temperature and humidity index are checked. The two methods in use include downgraded check and baseline check. For those regional natural and environmental conditions that have an impact on human life and production activities, the suitability of the identified space is downgraded based on the degree of impact; for the regional natural and environmental conditions that are not suitable for human activities thereof, the identified space is directly classified as unsuitable according to baseline check.

Safety risk factors check: The safety baseline space that poses potential risks to urban development is identified. This process includes identification of geological safety baseline characterized by major risks for urban construction, ecological security baseline marked by substantial risks for regional sustainability, and food security baseline indicative of significant risks for national development (Table 6.1).

Comprehensive carrying capacity check On the basis of the above evaluation results, the dynamic spatial scales model, such as the minimum resistance model, is used to simulate and check the comprehensive carrying capacity of the region under different scale scenarios, which helps to identify the upper threshold of spatial scale

Table 6.1 Safety risk baseline check

Name of factor	Risk check			Baseline check
	No impact	Downgraded by 1 degree	Downgraded by 2 degrees	Downgraded to unsuitable
Geological safety baseline	Others	The risk rating of ecological hazards are fairly high risk[7]	The risk rating of ecological hazards are high risk	The risk rating of ecological hazards are extremely high risk
Ecological security baseline	Others	/	/	Extremely important and sensitive ecosystem service areas
Food security baseline	Others	/	/	Basic farmland

Source Prepared by the author

in the region. The main targets of the simulation check include the carrying capacity of water resources and the environment.

Building upon the aforementioned steps, the evaluation results detailing the optimal development intensity for every district and county in China are ultimately crafted. By taking the county as the basic unit, the results are stratified based on the proportion of suitable development areas to the total county area and categorized into five grades utilizing the natural breakpoint method according to the distribution of results : ≤5%, 5–20%, 20–40%, 40–60%, and >60% .

(2) Precise identification of safety risk types

From the safety risk dimension, risk types of different regions are identified through three approaches: single-hazard risk assessment, disaster chain identification and delineation, and disaster type-based regionalization, to allow for precise identification of "safe" and "unsafe" spaces in national land space.

Single-hazard risk assessment Building upon the two dimensions of dangerousness of disaster-inducing hazards and vulnerability of exposures, a risk appraisal of single-hazard is undertaken for major disaster categories such as rainstorms, floods, earthquakes, landslides, typhoons, storm surges, hailstorms, snowstorms, extreme temperatures, and land subsidence. The dangerousness assessment of disaster-triggering hazards relies on historical disaster chronicles, genesis conditions, and changing patterns, combined with event frequencies of the past and forecasts to evaluate the severity of disaster events and their propensity of damage. Disaster vulnerability assessment pertains to evaluation of the system's susceptibility in response to triggers, including both material and social vulnerabilities.

[7] It includes earthquakes, landslides, avalanches, mudslides, subsidence, and other geological hazards.

Identification and delineation of disaster chain[8] By utilizing the probabilistic analysis method, the hierarchical clustering of major disaster types in China based on distribution of their dangerousness has revealed that snowstorm-hailstorm-low temperature, heavy rainstorm-flood, earthquake-thunderstorm-landslide have strong spatial and temporal alignment. They represent the types of disaster chains that affect the survival and development of cities in China.

Comprehensive disaster-type based regionalization Spatial clustering analysis for disaster-type based regionalization in China leads to identification of three major regions and nine sub-regions: hydrological disasters such as typhoons and intensive rainstorm-linked floods are prevalent in the eastern region; meteorological disasters such as low temperature-snowstorm-hailstorm disaster chain are primarily seen in the western and northeastern regions; and geological disasters characterized by earthquake-thunderstorm-landslide are mainly found in the northwestern and southwestern regions.

(3) Precise identification of cultural factors

To evaluate the distinctiveness of human settlements, factors that are selected include world cultural heritage, historical and cultural cities, towns, villages, neighborhoods and traditional settlements that can reflect the characteristics of the regional cultural landscape, as well as ethnicities, dialects, customs and architectural features that capture the cultural and social attributes. Based on the distribution of various cultural resources within the region by quantity and by magnitude of influence, a comprehensive evaluation index model is formulated. It evaluates the grades of different spaces in the region for their distinctiveness, in order to precisely identify the culturally-distinctive spaces that require strengthened preservation and oversight in national land space. Such cultural attributes by region are also identified and summarized.

6.5.2 Adaptive Distribution Technology

Adaptive distribution technology refers to the formulation of urban spatial patterns that are in harmony with nature and conducive to development, taking into account precise analysis, as well as population and economic factors specific to different regions. This process involves addressing two key questions:

Firstly, the question of "where to build cities." Firstly, based on precise analysis, suitable areas for construction can be identified. Secondly, from a developmental perspective, population and economic analyses are employed to determine where

[8] As practice has indicated, the unfolding of many natural disasters often instigates a cascade of secondary disasters. It is termed as disaster continuum or disaster chain. For research methodologies, the identification of disaster chains mainly involves probabilistic data-based analysis, complex network-informed analysis, and remote sensing-driven measurements.

people are likely to move, providing support for the establishment of city systems and the selection of locations for new urban areas.

Secondly, the question of "how to build cities." Based on the differentiated natural ecological conditions and safety risk types identified through precise analysis, the national land space is divided into various urbanization zones. This allows for the development of adaptive planning and construction models tailored to the specific conditions of different regions.

By integrating these technologies, a comprehensive spatial planning knowledge domains map is ultimately formed, covering multiple regions, scales, and disciplines.

1. Prediction of Urbanization Trends

Predicting the speed and level of urbanization nationwide and in different regions is crucial for assessing population migration trends and determining the direction of population movement. A nonlinear model is established to correlate urbanization with the level of economic development. Taking into account the characteristics of urbanization development in various countries worldwide, a comprehensive assessment and analysis of the stage and trend of urbanization in China is conducted. Regression analysis of the model reveals that the urbanization levels of different countries generally follow the sigmoid (S-shaped) curve, with variations in the starting point due to significant individual differences among nations. The relationship between China's urbanization process and the level of economic development also demonstrates an S-shaped curve, but in comparison to most countries with similar economic development levels, China's urbanization level at the same stage is relatively low, exhibiting distinct characteristics from those of typical large countries (Fig. 6.4).

Therefore, in the model analysis, "dummy variable" parameters reflecting China's national conditions and characteristics are adjusted to establish a more linear and

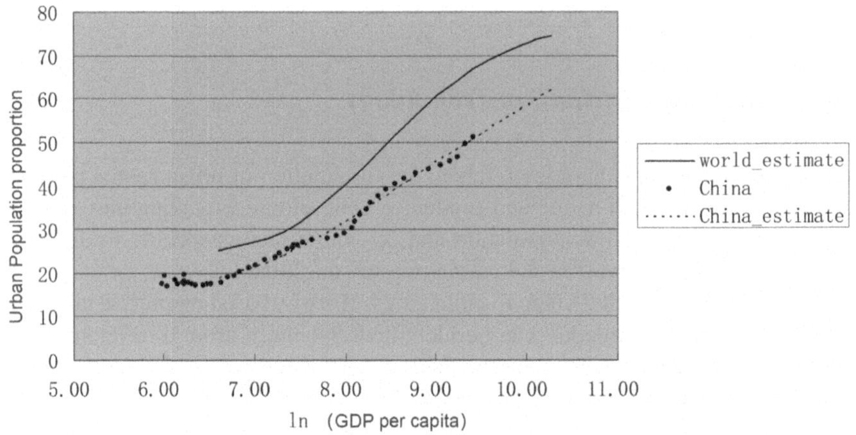

Fig. 6.4 Comparison chart of China's actual urbanization process and fitted curve. *Source Study on the Speed and Quality Evaluation and Forecast of China's Urbanization*, undertaken by the China Academy of Urban Planning and Design and chaired by the author

gently sloping "S" curve model specific to China's urbanization. Based on this, the development speed and trend of China's urbanization in the medium and later stages (key milestones in 2020 and 2030) are assessed: China's urbanization level is projected to reach approximately 59%-60% in 2020 and around 68%-70% in 2030. By utilizing panel data models with different regression coefficients based on the unique characteristics of each province, predictions for the speed and level of urbanization in different provincial regions of China are made.

2. The Construction of National Central City System

A national central city is a core city in the urban spatial system of national spatial planning. The construction of a national central city system and the determination of "where to build cities" are important aspects of urban spatial distribution in national spatial planning in China.

Characteristics and connotations of national central cities:

National central cities possess two fundamental characteristics at the national level: centrality and a certain level of internationality. In the context of globalization, the centrality of national central cities is reflected in their functions as national-level market and network centers, while their internationality is manifested through aspects such as international gateways and innovation centers. In the past decade, factors such as global climate change and the wave of technological innovation have propelled globalization into a new adjustment phase. Consequently, the connotations of national central cities have further expanded. On one hand, they demonstrate a high degree of integration between centrality and internationality, serving as a link in international connections and domestic network organization. On the other hand, as a populous country, new national central cities should consider both territorial balance and security support, ensuring a coordinated approach between security and development (Table 6.2).

The Construction of the national central city system:

Based on the connotations of national central cities, a measurement indicator system is constructed using scale-based indicators, network-based indicators, and potential-based indicators. The scale-based indicators encompass six aspects: international connectivity and international communication, regional public service capacity, innovation capacity and innovation environment, internal and external traffic organization and services, regional service capacity and level, and security and emergency response. To become a national central city, it is necessary for the number of indicators meeting the standard threshold to be no less than 80% of the total number of scale-based indicators (Table 6.3).

Network-based indicators are divided into two aspects: external linkages and domestic networks. A necessary condition to become a national central city is that the number of indicators meeting the standard threshold should not be lower than 85% of the total number of scale-based indicators.

Growth potential indicators are divided into four aspects: market potential, capability assurance, enterprise vitality, and innovation drive (Table 6.4).

Table 6.2 Scale-based indicators and standards for national central cities

Dimensions	Indicators	Thresholds
International connectivity and international communication (6 aspects)	Number of consulates or embassies (count)	≥10
	Number of international organization headquarters/representative offices (count)	≥5
	Number of fortune global 500 company headquarters and branches (count)	≥120
	Number of international cultural, sports, exhibition, and performance events (occurrences)	≥50
	Percentage of total goods and services import/export to GDP (%)	≥15
Regional public service capacity (4 aspects)	Number of grade III level a hospitals (including large medical centers and rescue centers) (count)	≥20
	Number of universities in the "985" and "211" project (count)	≥10
	Number of large-scale cultural and sports facilities (count, limited to specific scale)	≥10
	Number of government information disclosures (PB)	≥10
Innovation capacity and innovation environment (4 aspects)	Number of national high-tech enterprises per 10,000 people (count/10,000 people)	≥5
	Number of patents per 10,000 people (count/10,000 people)	≥60
	Number of national laboratories and key state laboratories (count)	≥10
	Percentage of total R&D expenditure to GDP (%)	≥2.8
Internal and external traffic organization and services (4 aspects)	Number of international air routes at airports (count)	≥20
	Number of domestic city-to-city flights at domestic airports (count)	≥50
	Air cargo throughput (10,000 tons)	≥200
	Number of high-speed railway (speed ≥200 km/h) origin and destination stations (count)	≥200
Security and emergency response (1 aspect)	Scale of emergency material reserves (100 million CNY)	√

Source Study on the Assessment Indicators for National and Regional Central Cities, a special study for urban health check in 2021 undertaken by the China Academy of Urban Planning and Design and chaired by the author

Table 6.3 Network-based indicators and standards for national central cities

Dimensions	Sub-dimensions	Indicators	Standards (Ranking)
External connectivity (3 aspects)	Global innovation network	Including innovation entities, innovative knowledge, and innovative services.	Global city ranking within the top 100.
	Global production and service network	Including global production networks and global service networks.	Global city ranking within the top 30.
	Global facility connectivity network	Including global aviation, intercontinental rail, and port transportation networks.	Global city ranking within the top 30.
Domestic networks (4 aspects)	Domestic innovation network	Including the transfer of scientific and technological achievements, citations of scientific papers.	Domestic city ranking within the top 20.
	Domestic production and service network	Including domestic production and service investment networks; logistics networks.	Domestic city ranking within the top 10.
		Including passenger service networks primarily composed of high-speed rail and aviation.	Domestic city ranking within the top 20.
	Passenger transportation network	Including various sources of information, information services.	Domestic city ranking within the top 20.
	Information network	Including innovation entities, innovative knowledge, and innovative services.	Domestic city ranking within the top 10.

Source Study on the Assessment Indicators for National and Regional Central Cities, a special study for urban health check in 2021 undertaken by the China Academy of Urban Planning and Design and chaired by the author

3. Urbanization Zoning

Precise analysis is used to define urbanization zones:

Defining urbanization zones and proposing differentiated construction models for different regions is one of the core components of national spatial planning.

Table 6.4 Growth potential indicators and standards for national central cities

Dimensions	Indicators	Evaluation methods
Market potential (1 indicator)	Population size	The entropy method is employed to determine the weights of each indicator, constructing a model for evaluating the economic development potential of cities. The model calculates the comprehensive scores of economic development potential for each city and categorizes them into three levels: "High Potential City," "Medium Potential City," and "Low Potential City," based on the following criteria: High Potential City: Comprehensive potential score ≥ 0.5 Medium Potential City: Comprehensive potential score ≥ 0.2 and < 0.5 Low Potential City: Comprehensive potential score < 0.2
Domestic network (1 indicator)	Total fixed asset investment in the entire society	
Enterprise vitality (2 indicator)	Number of IPO companies	
	Number of specialized, sophisticated, and innovative enterprises	
Innovation drive (3 indicator)	Number of industrial clusters	
	Population with college education or above	
	Industrial diversity	

Source Study on the Assessment Indicators for National and Regional Central Cities, a special study for urban health check in 2021 undertaken by the China Academy of Urban Planning and Design and chaired by the author

The policy zoning for urbanization, as outlined in Chapter 4 of the technical framework for the study of urban spatial development in China, is based on organizing regional economic activities according to economic zones, implementing national regional development strategies and urbanization policy requirements, and promoting comprehensive socio-economic development at a national level. In this section, urbanization zoning is based on precise analysis and identification of spatial resource baselines, building upon the policy zoning discussed earlier. It serves as an important tool and means for guiding specific planning and construction at the local level within the framework of national spatial planning.

Based on precise analysis and identification of the natural ecology and safety risks in different regions, combined with population and economic-related analyses such as urbanization trend predictions, identification of national central cities, and population density distribution, the national territory is divided into several urbanization zones. The zoning hierarchy includes primary zones and secondary zones. The primary zones reflect the overall differences in urbanization development and population distribution under different natural geographical conditions in China. These zones include ecologically sensitive and fragile areas unsuitable for construction, key urbanization areas, high-density population plain areas, low-density population plain areas, high-density population mountainous and hilly areas, and low-density population mountainous and hilly areas. The secondary zones, built upon the primary zones, reflect the variations in different regions in terms of safety risk types and resource and environmental carrying capacity (Table 6.5) (Fig. 6.5).

Table 6.5 List of urbanization zones in China

Primary zones		Description	Secondary zones	Risk degree	Disaster type description
1	Ecologically sensitive and fragile areas unsuitable for construction	Mostly located in ecologically sensitive and fragile areas with low population density, such as plateaus	1-a	Medium	Mostly medium-high risk in snow disaster chain. In certain areas, it involves high risk in both snow disaster and earthquake disaster chains
			1-b	Medium-low	Mostly not high-risk areas for disasters, with certain areas of medium-high risk in the earthquake disaster chain
			1-c	Medium-low	Mostly not high-risk areas for disasters, with certain areas of medium-high risk in the snow disaster chain
2	Key urbanization areas (major urban clusters, metropolitan coordinating areas, central cities)	Mostly located in important urban clusters, metropolitan coordinating areas and central cities, with high population density	2-a	High	Mostly involves high risk in three disaster chains, with some high-risk areas in typhoon and snow disaster chains. Accompanied by water scarcity
			2-b	High	Mostly involves high risk in earthquake and typhoon disaster chains, with some high-risk areas in three disaster chains. Accompanied by water scarcity in northern regions

(continued)

Table 6.5 (continued)

Primary zones		Description	Secondary zones	Risk degree	Disaster type description
			2-c	Medium-high	Mostly involves high risk in typhoon and snow disaster chains or medium-high risk in typhoon disaster chain. Accompanied by water scarcity in northern regions
			2-d	Medium	Mostly involves medium-high risk in snow disaster chain, with some high risk in earthquake and snow disaster chains. Accompanied by relative water scarcity
			2-e	Medium-low	Not in high risk, with some medium-high risk in earthquake disaster chain. Accompanied by water scarcity (except for Xining and Lhasa)
3	Plain areas with high population density	Plain areas or gentle plateau areas with high population density	3-a	High	Mostly involves high risks for all three disaster chain. Accompanied by water scarcity
			3-b	Medium-high	Mostly involves high risks for earthquake and typhoon disaster chains. Accompanied by water scarcity

(continued)

Table 6.5 (continued)

Primary zones		Description	Secondary zones	Risk degree	Disaster type description
			3-c	Medium-high	Mostly involves high risks for typhoon and snow disaster chains or medium-high risk for snow disaster chain. Accompanied by water scarcity
			3-d	Low	Not high risk for disasters, with certain areas of medium-high risk for earthquake disaster chain. Accompanied by relative water scarcity
4	Plain areas with low population density	Plain areas or gentle plateau areas with low population density	4-a	Medium	Mostly involves medium-high risk for snow disaster chain
			4-b	Low	Not high risk for disasters
5	Mountainous and hilly areas with high population density	Mountainous and hilly areas with relatively high population density	5-a	Medium-high	Mostly involves high risks for earthquake and typhoon disaster chains, with certain areas of high risk for all three disaster chains, featuring sensitive and vulnerable ecology, accompanied by water scarcity
			5-b	Medium-high	Mostly involves high risks for typhoon and snow disaster chains. Accompanied by water scarcity

(continued)

Table 6.5 (continued)

Primary zones		Description	Secondary zones	Risk degree	Disaster type description
			5-c	Medium	Mostly involves medium high risks for typhoon or high risks for typhoon and snow disaster chains, featuring sensitive and vulnerable ecology
			5-d	Medium	Mostly involves medium high risks for typhoon or high risks for typhoon and snow disaster chains, featuring sensitive and vulnerable ecology
6	Mountainous and hilly areas with low population density	Mountainous and hilly areas with relatively low population density	6-a	Medium-high	Mostly involves high risks for earthquake and typhoon disaster chains or medium-high risks for typhoon, with certain areas of high risks for all three disaster chains, featuring sensitive and vulnerable ecology
			6-b	Medium-high	Mostly involves high risks for earthquake and snow disaster chains. Accompanied by water scarcity

(continued)

Table 6.5 (continued)

Primary zones		Description	Secondary zones	Risk degree	Disaster type description
			6-c	Medium	Mostly involves medium-high risks for snow disaster chains, featuring sensitive and vulnerable ecology
			6-d	Medium-low	Not high risk for disasters in most areas, with certain areas of medium-high risk for earthquake disaster chain, featuring sensitive and vulnerable ecology. Accompanied by relative water scarcity in the north

Source Prepared by the author

Propose adaptive urban construction models:

Based on the delineation of urbanization zones, differentiated and adaptive urban construction models are proposed according to the variations in natural environmental conditions and types of safety risks in different regions. For example, in mountainous and hilly areas prone to geological hazards such as earthquakes, landslides, and debris flows, large-scale concentrated urban construction should be avoided, and cluster-based distribution patterns that are compatible with the terrain and geological conditions should be adopted. To address earthquake disasters, strict control over building density and intensity should be enforced, and appropriate seismic design standards and construction techniques for buildings and engineering facilities should be developed.

It is worth noting that, against the backdrop of intensified global climate change, special attention should be given to urbanization key areas and urbanization special areas characterized by extreme climate conditions such as high cold, high heat, and high altitude. The high-intensity and high-density development issues in urbanization key areas, as well as the enormous losses caused by public health and human-induced accidents, are global challenges. Therefore, the planning industry needs to systematically explore adaptive strategies at the regional-city-block levels based on safety concerns. Urbanization special areas are strategic regions for national security, requiring tailored planning techniques for safety construction. For instance,

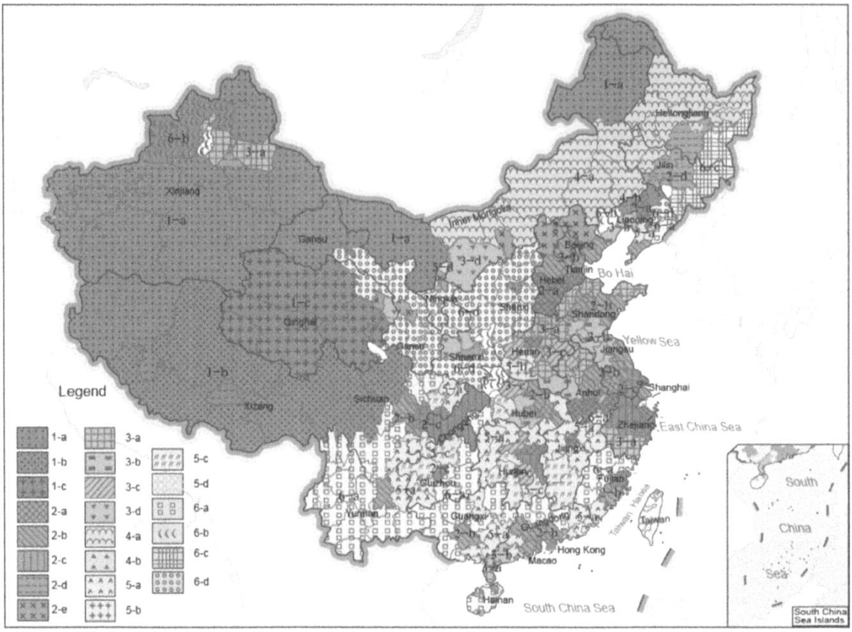

Fig. 6.5 National urbanization policy zoning. *Source* Planning and Design and chaired by the author

in high-altitude areas, scientific site selection and layout are needed, along with enhanced analysis of land suitability, ecological sensitivity, and low-impact planning and construction techniques. In high cold areas, scientific experimental bases should be established to build secure living environments adapted to the site. In high humid and hot areas, such as China's South China Sea, a regionally linked service support system should be constructed based on island spatial distance and unique transportation methods, and the impacts of humidity and heat should be thoroughly considered in base construction.

4. Knowledge domains map of spatial planning

Drawing upon years of research and planning practice, an adaptive knowledge domains map has been distilled to address differentiated construction models for various regions, guiding planning and construction in different areas. It is important to note that the methods of precise regional analysis and adaptive technology are not only applicable to macro-scale spatial planning at the national and regional levels but also play a crucial role in planning and construction techniques at different scales such as urban areas, neighborhoods, buildings, and engineering facilities. For example, based on the identification of regional safety risks, the knowledge domains map analyzes and determines the development intensity, construction scale, infrastructure engineering, and building standards and techniques of cities. Therefore, the planning and construction key points of this knowledge domains map encompass multiple scales, from macro to micro. Ultimately, it integrates planning and construction techniques across multiple scales and disciplines, encompassing regional planning, urban planning, architecture, and engineering facilities, from a macro-regional perspective (Fig. 6.6).

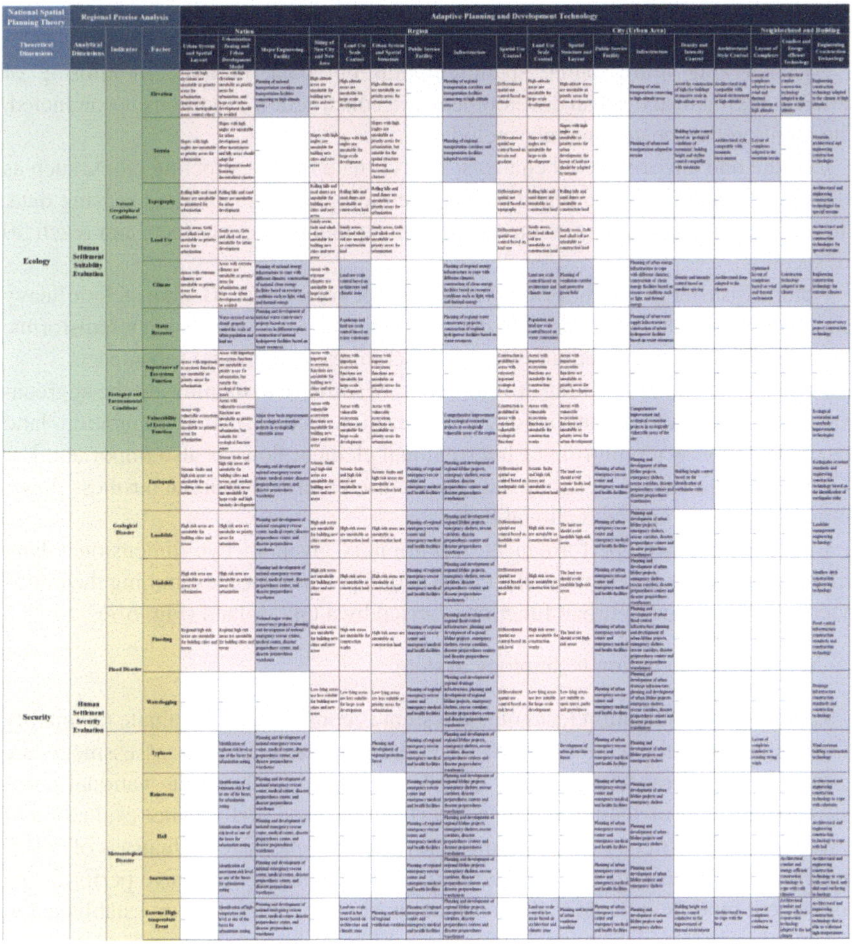

Fig. 6.6 Knowledge domains map of adaptive technology. *Source* Prepared by the author based on the practices of several projects

6.5.3 Dynamic Evaluation Technology

The implementation of real-time monitoring, periodic evaluation, and timely adjustment and optimization is an integral part of a comprehensive planning system that addresses the uncertainty of external changes throughout the entire planning cycle.

1. Dynamic Monitoring

By analyzing multi-source spatiotemporal big data, periodic monitoring of national, regional, and urban development and construction conditions is conducted. The monitored indicators mainly include:

Climate and environmental changes: Key climate factors affecting cities, such as temperature, humidity, and precipitation, are monitored using remote sensing data. The monitoring accuracy of mainstream climate monitoring products can reach 30 arc seconds.

Natural disaster risks: The primary disaster types of interest in China include heavy rain, floods, earthquakes, landslides, debris flows, typhoons, storm surges, hailstorms, snow disasters, high temperatures, low temperatures, and ground subsidence.

Population and urbanization development: In addition to monitoring the aggregation trends and development trends of urbanization factors such as population, land use, and economy, monitoring of interregional element flows is also important. For example, mobile phone signaling data can be used to monitor urban business flows, commuting flows, and recreational flows.

Land development performance: Attention is given to two-dimensional land performance and three-dimensional construction performance, examining the multifaceted performance of urban functions carried by a unit of land (Fig. 6.7).

2. Multi-level evaluation technology system

A multi-level evaluation technology system is constructed to evaluate the quality of urbanization at the national, provincial, and city levels. Specific assessments are conducted to address key issues at each level. For example, at the national level, assessments focus on land development performance and performance evaluation of new towns and districts. At the provincial level, comprehensive evaluations of development zones are conducted. At the regional level, spatially intensive performance assessment models and regional coordination assessment models are established to analyze the characteristics and efficiency of different factors, such as land, population, and economy, along with their synergistic relationship with the environment, in order to identify development shortcomings in the region. At the city level, AI analysis techniques utilizing large sample data are employed. Diagnostic methods such as spatial correlation analysis of indicators, spatial multidimensional clustering based on differences, and effective coverage based on supply and demand matching are used to identify strengths and weaknesses in urban planning and construction.

3. Dynamic extrapolation techniques

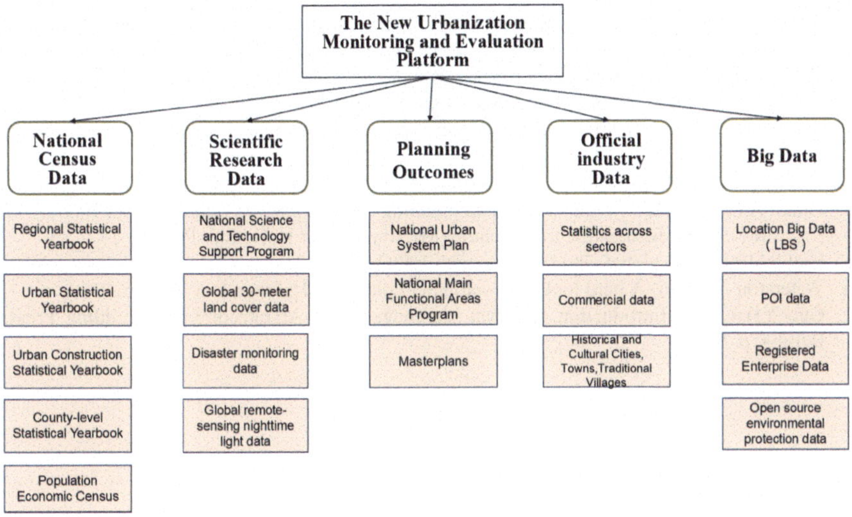

Fig. 6.7 Data directory of the new urbanization monitoring and evaluation platform. *Source Research and Development of National Urban and Rural Planning Information Platform (2017-2019)*, a project undertaken by the China Academy of Urban Planning and Design and chaired by the author

National spatial planning requires the assessment of spatial development trends over a relatively long period, necessitating the extrapolation of monitoring indicators related to long-term factors. For instance, indicators such as annual average precipitation and extreme precipitation changes resulting from global climate change can be dynamically extrapolated through long-term monitoring. Similarly, the distribution of population and economy needs to be accurately understood over extended periods. Therefore, dynamic extrapolation of these long-term monitoring indicators can provide predictive support for the overall and systematic optimization and adjustment of national territorial space.

References

1. Hu WJ (2005) The political logic of equilibrium. Chongqing: Chongqing Publishing House 18:46, 50.
2. Castells M (1989) The informational city: information technology, economic restructuring, and the urban-regional process[M]. Blackwell, Oxford
3. Amin A, Thrift N (eds) Globalization, institutions, and regional development in Europe, Oxford University Press, 1994.
4. Morgen S (2001) The agency of welfare workers: negotiating devolution, privatization, and the meaning of selfsufficiency[J]. Am Anthropol 103(3):747–761
5. Cao PL (1998) Government and market. Zhejiang People's Publishing House, Hangzhou, p 6
6. Hall P (2002) Urban and regional planning, 4th edn. Routledge, London
7. Hall S (2002) The yangtze river Paradigm. City Plan Rev (12)

8. Alden DL ,Steenkamp JBEM, Batra R (1999) Brand positioning through advertising in Asia, North America, and Europe: the role of global consumer culture[J]. J Mark 63(1):75–87
9. Hall P (1996) Cities of tomorrow, Blackwell
10. Friedmann J (1986) The world city hypothesis[J]. Dev Change 17(1):69–83
11. Friedman J (2005) Life space and economic space: contradictions in regional development. Urban Plan Int 5:8
12. Gu CL et al (2002) City governance. Southeast University Press, Nanjing, p 246
13. Faludi A (1973) The "systems view" and planning theory[J]. Socio-Econ Plan Sci 7(1):67–77
14. Evans R (1997) Regenerating town centres[M]. Manchester University Press
15. Hall P (1996) The global city[J]. Int Soc Sci J 48(147):15–23
16. Waterston A (1966) A hard look at development planning[J]. Finan Dev 3(2):85–91
17. Guo YH (1992) Introduction to urban planning. China Architecture and Building Press, Beijing, p 2

Chapter 7
Conclusions and Prospects

7.1 Main Conclusions

Over the past 40 years of reform and opening up, China's macro-level spatial planning has experienced periods of growth, decline, and resurgence. The rise and subsequent abandonment of national land planning in the mid-1980s, the silence throughout the 1990s, the subsequent focus on national-level spatial planning by various ministries in the early twenty-first century, and the formulation of the National Land Spatial Planning Outline under the State Council's institutional reform in 2018 reflect the evolving understanding of the central government's macroeconomic regulation in response to the establishment of a socialist market economy and the continuous advancement of urbanization in China.

In this era of significant changes, there is an urgent need for innovative theoretical and methodological support that is tailored to China's context. Previous research on regional planning and urban system planning has primarily focused on economic development as the core objective and the layout of productive forces as the main content. However, these approaches have struggled to address issues such as resource and environmental depletion, frequent natural disasters, and the erosion of regional cultural characteristics that have arisen during China's rapid and large-scale urbanization process since the reform and opening-up. Over the past four decades, the author has been engaged in regional planning research, conducting comprehensive studies on macro-level urban spatial planning from theoretical to practical perspectives, and from historical analysis to future considerations. As a result, the author has developed the theoretical approach and technical system of "National Spatial Planning Theory." The main conclusions are as follows:

1. The development of national urban space is directly influenced by political, economic systems, and policies. Planning serves as a guiding and regulatory tool for the government and is an integral part of national development strategies.

K. Wang, *National Spatial Planning in China*, Urban Sustainability,
https://doi.org/10.1007/978-981-97-7729-7_7

The emergence and development of urban space are inherently linked to the political and economic situations at that time and the conditions of nature, geography, and the environment. In China's industrialization and urbanization processes, proper ideological guidance and rational planning measures can significantly enhance the country's economic and social benefits, while incorrect approaches can lead to severe losses. Urban spatial planning plays an irreplaceable role in the reconstruction of urban spatial structures. It is not merely a technical process but also a governmental and social act. It is instrumental in improving national competitiveness and serves as a crucial means to prevent and rectify market failures. From an administrative perspective, national urban spatial planning serves as a general principle for coordinating interests among various government departments, the central government, and local governments. It plays a significant role in promoting the common development of different regions and preventing the excessive amplification of market forces.

2. The scientific and rational development of China's urban space must be based on the principles of the science of human settlements, with a theoretical and methodological framework that embraces a global perspective, resource constraints, dynamic development, equitable service provision, and sustainable development.

The development of China's urban space must be closely integrated with global economic and social development, enabling different regions to integrate into the world's economic tide and construct an urban spatial structure that is oriented towards and integrated into the world. Emphasizing the constraints imposed by resources and the ecological environment on spatial development, precious resources that impact national interests and maintain overall habitat quality must be protected. A comprehensive consideration of the five systems of nature, people, society, living, and support networks is required to construct an ecologically safe, economically efficient, and harmonious urban–rural spatial system. Faced with the continuous changes in natural, economic, and social environments, an open and flexible urban spatial structure must be established to enhance its adaptability. Additionally, the service level of regional transportation infrastructure must be improved to facilitate spatial growth. The principle of equitable service provision should guide the establishment of a unified standard system for municipal and social infrastructure, promoting urban–rural integration. The complex relationships between the central and local governments and among different local governments must be managed effectively. Intervention policies for spatial development at different levels should be determined to ensure the coordinated development of the economy, society, and environment. In response to the challenges posed by globalization, informatization, marketization, as well as increasing security risks resulting from climate change, a stronger analysis of security risks is necessary to construct a more balanced, secure, and sustainable national urban spatial pattern.

3. The spatial distribution of urban areas must be closely integrated with the social and economic development strategies, and the national strategies for social and economic development also need to be coordinated with urban development.

Throughout the four historical periods of urban spatial development in China since its founding, the development and planning of urban space have been directly influenced by political and economic systems. Its occurrence and development are inseparable from the political and economic situations at that time and in specific locations. Correct planning guidance leads to significant economic and social benefits for the country, while incorrect guidance can result in severe losses. The success of 156 key projects is attributed to the coordination between productive spatial distribution and urban development, while the failure of the "Third Front" construction was due to its detachment from the context of urban development. These positive and negative cases provide us with profound experiences and lessons. During more than 40 years of reform and development, the sustained rapid economic growth in China can largely be attributed to the determination of a national spatial strategy centered on coastal openness and urbanization. The enduring relevance of Dr. Sun Yat-sen's *The Strategies for Founding A Country* up until now is attributable to its guidance from spatial planning.

4. The theoretical approach and technical system of "National Spatial Planning Theory" and the "Precise Analysis-Adaptive Technology-Dynamic Evaluation" have provided important guidance and scientific support for the rational distribution of national and regional urban space under the complex national conditions in China.

Achieving sustainable urbanization in the complex context of our country poses an unprecedented and daunting challenge for the Chinese nation and humanity as a whole. In the face of this challenge, national and regional urban spatial planning should go beyond the limitations of previous regional planning that focused on production layout and land planning that emphasized land resource conservation and utilization. Instead, it should be built upon the basic principles established by the science of human habitat environment, taking into account a broader range of complex factors such as resource environment, population, industry, and institutions to ensure comprehensive and coordinated planning. The "National Spatial Planning Theory," with its core concept of "precise adaptation," proposes an approach based on the analysis of natural and cultural spatial resources, combined with analysis of national development strategies related to population and economy. It establishes a technical system of precise analysis, adaptive techniques, and dynamic evaluation, providing an effective theoretical and methodological framework for constructing urban spatial patterns that are compatible with nature and culture, while considering development. This approach facilitates the overall optimization of national territorial space.

7.2 Future Prospects

Nowadays, globalization has entered a new phase of adjustment, with increased security risks in industrial and supply chains, disruptive changes accelerated by informatization, and a rise in extreme weather events caused by global climate change. These factors have unprecedentedly stressed natural ecosystems, urban economic systems, and human societies. Looking ahead, for a considerable period of time, the optimization of China's urbanization spatial pattern will face numerous security risks and challenges, necessitating focused attention on the following three aspects:

Firstly, it is essential to integrate the spatial distribution of national innovation resources with the enhancement of technological originality, revitalization of old industrial bases, promotion of cross-sector integration innovation, and support for border area development. By effectively implementing the role of "government intervention," a more balanced development of national land space can be achieved.

Secondly, from the perspectives of security and sustainability, particular attention should be given to key areas of urbanization and areas with specific urbanization characteristics. High-intensity and high-density development in key urbanization areas, as well as significant losses caused by public health and human-induced accidents, are global challenges. Areas with specific urbanization characteristics, such as high-altitude, high-cold, and high-humid-hot regions, are strategic security areas for the nation, requiring adaptive safety and resilience planning techniques tailored to their specific circumstances.

Thirdly, robust strategies and adaptive management should be implemented in response to climate change. The intertwining of economic development, urbanization, and climate change necessitates urgent risk management. In the context of future changes in population, technology, market, and climate, decisions must be made based on incomplete information, with robustness serving as a decision-making criterion.

In summary, China's urbanization development has faced different contradictions and issues at different stages, and the adjustment of the national urban spatial pattern has also gone through various phases. From a focus on productivity spatial distribution to an emphasis on ecological priority and subsequently addressing security and cultural concerns, the theoretical approaches and technological advancements in national spatial planning require continuous exploration and improvement to promote China's distinctive and steadfast journey of urbanization.

Appendix
Outcomes Assessment of the Research Report on National Urban System Planning (2006–2020) (The Study Is a Report on the Scientific and Technical Results Based on the National Urban System Planning Outline (2005–2020) Organized By the Ministry of Construction, and Some of the Conclusions of the Study Have Been Adjusted and Supplemented)

In 2005, chaired and organized by the author, a consortium of research institutions, including the China Academy of Urban Planning and Design, the Institute of Geographic Sciences and Natural Resources Research of the Chinese Academy of Sciences, and Renmin University collaboratively prepared *The Research Report on National Urban System Planning (2006–2020)*. Grounded in the national realities of large population, strained per capita resource endowment, vast territory, and unbalanced regional development in China, alongside prominent problems such as waste of land resources, ecological deterioration, cultural and historical heritage degradation, and emergence of urban malaise, the study crafted a strategy to foster healthy urbanization characterized by an intensive, compact, and green approach. It reinforces the strategic allocation of spatial resources at the national level and baseline controls.[1] It serves as a robust empirical foundation for continuous enrichment and improvement of the *National Spatial Planning* based on "precision adaptation". Through nearly two decades of scrutiny, it has been demonstrated that the national and regional planning informed by this theory is aligned with China's development realities. It has provided a scientific bedrock for shaping the spatial distribution of China's urbanization over the past two decades, providing effective guidance for the formulation of major inter-regional plannings such as those for the Beijing-Tianjin-Hebei Region, the Yangtze River Delta, the Guangdong-Hong Kong-Macao Greater Bay Area, and the Chengdu-Chongqing economic circle.

[1] As articulated in the entry of "National Urban System Planning" in Encyclopedia of China, compiled by Wang Kai et al.

K. Wang, *National Spatial Planning in China*, Urban Sustainability,
https://doi.org/10.1007/978-981-97-7729-7

Summary of The Research Report on National Urban System Planning (2006–2020)

1. Urbanization development policy

With commitment to an urban development strategy underscored by "urban-rural integration, regional coordination, well-established functions, intensive development, social harmony, safety and efficiency, cultural prosperity, and pleasant human settlement".

2. Urbanization development targets

By 2020, China envisages a total population of approximately 1.45 billion, with urbanization rate ranging from 56% to 58%, and an urban populace tallying between 810 million and 840 million residents. The average annual growth rate of urban construction land is to be reined in at below 3%, ensuring that its total area is kept under 110,800 square kilometers by 2020.

3. National urbanization strategy

Building upon the characteristics and developmental trajectories of different regions, the planning activity provides guidance for varying regions and cities to pursue tailored urban and regional development through guiding measures such as urbanization policy zoning, spatial organization of the urban system, and urban development guidelines for priority regions and by provinces. Additionally, it enhances the extent of intensiveness and conservation in urban development through mandatory benchmarks for urban construction land use, water conservation, energy consumption metrics, atmospheric and water quality standards compliance, solid waste disposal, among others for higher quality urbanization.

4. Urban Spatial Organization and Distribution

The plan centers around city clusters at its core, supported by major connecting avenues as its backbone to facilitate regional cooperation in urban areas and important central cities as its focal hubs, to generate a diversified, multi-polar and network-based urban spatial pattern. Diversity entails tailored spatial organization and developmental models for cities and towns based on local conditions, such as resource availability, developmental stages, mechanisms, and regional typology. Multi-polarity leverages diverse city clusters and central cities across varying tiers and types to foster regional growth and implement the overarching strategy of coordinated regional development. Network-based interconnectivity involves formation of close ties, strengths complementarity, and unimpeded factor flows across central cities and in between cities and urban-rural areas based on transportation corridors. Such network-based urban spatial system features strong interconnectivity between cities of varying sizes and small townships, reasonable distribution and coordinated development, with city clusters and various tiers of central cities at its core. It builds upon major national land passages, waterways and coastlines (Fig. A.1).

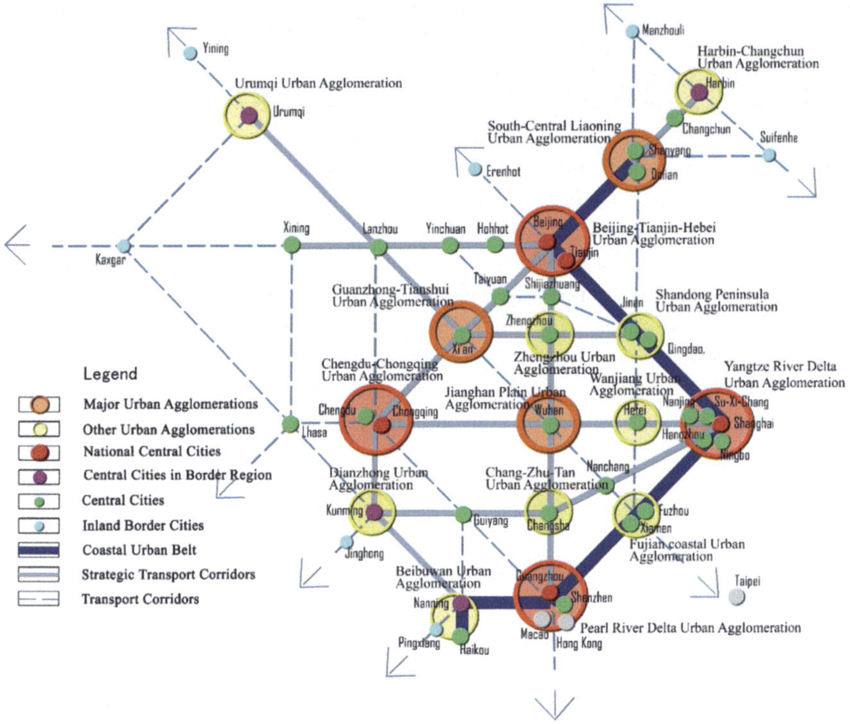

Fig. A.1 Diagram of China's National Urban Spatial Distribution (2020). *Source* The Research Report on National Urban System Planning (2006–2020)

5. Urban classification and development strategy

Emphasizing its commitment to an urban development strategy of "urban-rural integration, regional coordination, well-established functions, intensive development, social harmony, safety and efficiency, cultural prosperity, and pleasant human settlement", the plan shepherds reasonable urban development across different regions, typologies, and scales in an orderly manner. The national urban system planning has strengthened classification-based guidance for central cities, land gateway cities/townships, borderland central cities, cities with old industrial bases, mineral/resource-rich cities and historic and cultural cities.

National-level guidelines for priority regions. Priority regions include city clusters, cross-provincial coordinated urban zones, major river basins, lake districts, and coastal areas. Such key regions hold strategic significance in increasing national competitiveness, as well as coordinating regional development, and resource conservation. The state is committed to bolstering policy frameworks that govern urbanization and urban development management within these priority regions.

6. Distribution of national integrated transportation hubs

Following the energy strategy, resource conservation, and transportation security principles, a nationwide integrated transportation hub system is created. This system will amplify the pivotal role of cities and towns in industrial development and spatial distribution, fostering seamless intermodal connectivity while enhancing central cities' capacity for better urban access. Centering around the national integrated transportation hub system with central cities as key nodes, an integrated transport modality is rolled out to enhance connectivity between various transport modes and hub infrastructure. It allows for seamless and distance-friendly passenger interchange and cargo transfer, with better delivery of transportation services in gateway cities and strengthened core role of central cities in industrial development and spatial distribution. This initiative strives to foster intra-city and inter-regional transport seamlessly and create an efficient, convenient, equitable, and orderly urban transport system.

Outcomes Assessment of Research Implementation

Remarkable improvements in the competitiveness of national priority regions of urbanization. The diversified, multi-polar and network-based urban spatial pattern proposed by this report has basically materialized. The five major city clusters have already taken shape and grown into core regions that pioneer nation-wide development, namely Beijing-Tianjin-Hebei Region, Yangtze River Delta, Pearl River Delta (Guangdong-Hong Kong-Macao Greater Bay Area), the middle reaches of the Yangtze River, and Chengdu-Chongqing Economic Circle. Such clusters stand out for their pronounced global influence in innovation strengths with their wealth of innovation resources. By 2021, these clusters collectively host a staggering 93.6% of the China's patents and 97% of its high-tech enterprises. Notably, Beijing-Tianjin-Hebei Region, the Yangtze River Delta, and the Pearl River Delta stand out as prime convergence points, accommodating 88.1% of unicorns and 76.9% of gazelles. Meanwhile, the three major city clusters have seen the most pronounced translation of innovations into results among them. The coastal city corridor proposed by the report has already taken shape, with high density of cities, huge size of population and dense distribution of national transportation corridors. Large-scale airports and seaports are being built.

The blueprint for integrated transportation hub cities at the national level has materialized, alongside implementation of strategic planning on national transportation corridors. The Ministry of Transport in China, as outlined in the 13th Five-Year Plan, has committed to orchestrating integrated transportation networks and hubs across the nation, based on the report's core tenets. The period from 2015 to 2023 has seen an investment exceeding CNY 28 trillion (USD 3.9 trillion)in fixed assets for high-speed rails, expressways, ports, and airports along major transportation routes. The report proposes nine major integrated transportation hub cities, with Beijing,

Shanghai, and Chengdu designed to be equipped with two 4E airports. Noteworthy is the transformation of Shanghai Hongqiao hub into a pivotal link and engine for economic growth in the Yangtze River Delta. The proportion of business passengers in the air-rail passenger traffic of Hongqiao hub increased from 36% in 2012 to 51% in 2022. Zhengzhou, as the biggest beneficiary of national-level hub integration, has witnessed remarkable growth in its aviation hub, spurring the development of an airport economic zone for close to 100 square kilometers and cargo volume increased from 250,000 tons in 2013 to the maximum of 705,000 tons in 2021.

The efficiency of intensive urban space utilization in China has registered an upswing. By 2022, urban construction land spans 129,000 square kilometers,[2] constituting 1.3% of the nation's land area. It marks a noteworthy increase over the 78,000 square kilometers in 2005, with a 0.5 percentage point increase in proportion. The intensity of land development and the integrated performance of urban clusters show commendable results,[3] especially in key regions like the Yangtze River Delta, Pearl River Delta, and Chengdu-Chongqing urban cluster. Noteworthy progress in land development intensiveness is seen in recent years. The per capita construction land index of China's 21 super and mega cities in 2020 is 105 square meters, being kept within the reasonable regulation range in the land use national standard. It comes down by about 4 square meters from 2010, equivalent to a total of 1.05 million mu of land saved.

The planning and implementation of subsequent regional strategies have been bolstered. A decade after finishing *The Research Report on National Urban System Planning (2006–2020)*, the China Academy of Urban Planning and Design spearheaded the preparation of the National Urban System Planning Report for 2015–2030. This initiative has catalyzed enhancements in national development and conservation pattern, fostering holistic and coordinated resource utilization and cross-regional coordination endeavors. Since then, regional planning initiatives in China have evolved from economic growth-centeed orientations to integrated approaches with multiple targets, emphasizing safety, sustainability, livability, culture, and innovation. Exemplary regional schemes include the Beijing-Tianjin-Hebei Urban and Rural Planning (2016–2030) and the National Spatial Master Plan for Yangtze River Delta Ecological, Green and Integrated Development Demonstration Zone (2021–2035). The former focuses on the coordinated spatial development of the Beijing-Tianjin-Hebei city cluster as a whole under resource constraints; the latter coordinates the ecological security of the bordering areas with the high-quality development of urban and rural areas.

---

[2] According to the China Urban–Rural Construction Statistical Yearbook 2005 and China Urban–Rural Construction Statistical Yearbook 2022, the urban construction land includes three components: those of cities, counties and other designated administrative towns.

[3] The integrated spatial performance refers to the weighted performance of city's residential, industrial, service and transportation functions in an integrated manner.

The manufacturer's authorised representative in the EU is Springer Nature Customer Service Centre GmbH, Europaplatz 3, 69115 Heidelberg, Germany. If you have any concerns regarding our products, please contact ProductSafety@springernature.com

Printed and bound by CPI Group (UK) Ltd, Croydon, CR0 4YY
01/10/2026
02225858-0007